Nonlinear Mathematics and its Applications

NONLINEAR MATHEMATICS AND ITS APPLICATIONS

edited by

Philip J. Aston

University of Surrey

CAMBRIDGE
UNIVERSITY PRESS

University Printing House, Cambridge CB2 8BS, United Kingdom

Cambridge University Press is part of the University of Cambridge.

It furthers the University's mission by disseminating knowledge in the pursuit of
education, learning and research at the highest international levels of excellence.

www.cambridge.org
Information on this title: www.cambridge.org/9780521576765

© Cambridge University Press 1996

First published 1996

A catalogue record for this publication is available from the British Library

ISBN 978-0-521-57190-6 Hardback
ISBN 978-0-521-57676-5 Paperback

Contents

Preface

A Spring School in Applied Nonlinear Mathematics for postgraduate students was held at the University of Surrey on 3–7 April 1995, supported by EPSRC. This was the fourth such Spring School, following on from previous events at the University of Cambridge in 1991, the University of Bath in 1992 and the University of Leeds in 1994. The School was attended by 55 students who came not only from Mathematics Departments, but also from Engineering and Physics Departments.

The talks given at the School all demonstrated the way in which nonlinear mathematics can be used to understand problems arising in areas such as engineering, fluid mechanics, materials science and biology. This written version of the material presented provides a coherent account of recent research in these areas thus making this volume an ideal introduction into a variety of research fields. Furthermore, each chapter contains an extensive list of references which provides an opening into the research literature on each topic. Students embarking on mathematical research will benefit from the exposition on a variety of topics, while many from other disciplines will welcome this material as a means of gaining some understanding of areas which interest them but which they find hard to get into.

I am grateful to EPSRC for their financial support and to Prof J. Brindley for his support and encouragement in the arranging of the School. Dr J. Ockendons contribution to the School was appreciated, although it is not included in this volume. I would also like to thank Dr T. Bridges, Dr M.V. Bartuccelli, Dr S. Gourley and Mr C. Bird for their help in running the School and making it such a success.

Philip Aston
University of Surrey
February 1996

Global Dynamics of Driven Oscillators: Fractal Basins and Indeterminate Bifurcations

J.M.T. Thompson

Centre for Nonlinear Dynamics and its Applications,
University College London,
London

1 Nonlinear Dynamics and Chaos

It is remarkable that some 300 years after the publication of Newton's Principia, in 1687, we are today witnessing a Renaissance in the general field of dynamics. Pioneering studies of the topology of phase space by Poincaré (1854-1912) triggered the mathematical developments of Birkhoff, Smale, Arnol'd, and others. This rather abstract body of work was brought dramatically to life with the advent of computers in the second half of this century. In particular the computational studies of Lorenz and Ueda identified chaotic solutions as robust and typical features in the response of simple low-dimensional sets of ordinary differential equations. Before we proceed with our detailed examination of driven oscillators, we start with a brief outline of nonlinear dynamics and chaos which will serve to introduce the reader to some of the specialised concepts involved. Useful introductions to the subject are Abraham & Shaw [1], Arrowsmith & Place [2], Glendinning [9], Guckenheimer & Holmes [12], Moon [25], Parker & Chua [27], Thompson [42] and Thompson & Stewart [52, 53]. Recent collections of applications are Kim & Stringer [13], Mullin [26], Schiehlen [29], Thompson & Chua [43], Thompson & Gray [44] and Thompson & Schiehlen [48].

1.1 Dynamical systems

Dynamical systems can be classified as either continuous or discrete. A continuous system is one in which a set of differential equations describes the evolution of the system in continuous time. A discrete system is governed by an iterated map which describes the evolution of the system in discrete time. Dynamical systems can also be classified as either conservative (Hamiltonian) or dissipative. Conservative systems are relevant to celestial and particle physics. Dissipative systems, involving some form of energy dissipation or damping, arise in the modelling of the macroscopic systems encountered in engineering and applied physics. They differ fundamentally from conservative systems by exhibiting transients leading to asymptotically stable states.

In this article we concentrate on dissipative continuous systems, described by a finite set of ordinary differential equations. Without loss of generality we assume these to be in autonomous first-order form, with the time-dependent variables defining a multi-dimensional phase space. In this space a stationary vector field defines the non-crossing trajectories of the system. A one-degree-of-freedom driven oscillator can for example be cast in this form, with a 3-dimensional phase space spanned by the displacement x, the velocity y and the time t. Here the time t can be formally introduced as a phase variable by writing the dummy equation, $dt/dt = 1$. Notice that a Poincaré section will often be used to reduce a continuous dynamical system to a discrete mapping.

1.2   Attractors

After an initial transient response, the motion of a dissipative system settles onto an attractor, which can be of one of the following forms: a point attractor, a periodic attractor, a quasiperiodic attractor, or a chaotic attractor.

An example of a point attractor is the vertical hanging equilibrium state of a damped and unforced pendulum. All typical motions of the pendulum are attracted to this stable state, which corresponds to a fixed point in the 2D phase space spanned by the angle of the pendulum and its time derivative. An example of a periodic attractor is the stable resonant oscillation of a lightly driven pendulum under direct (horizontal) excitation. Since this has the same frequency as the sinusoidal driving it is referred to as a harmonic (as opposed to a subharmonic) oscillation of "period-one", written briefly as $n = 1$. This oscillation can be viewed as a closed orbit in the 3D toroidal phase space defined by position, velocity and forcing phase. Under stronger driving of a pendulum, stable subharmonic motions can arise consisting of periodic attractors with periods that are multiples, n, of the driving period. Quasiperiodic attractors, containing two or more incommensurate frequencies, arise typically in driven oscillators of the van der Pol type and in the escape from higher dimensional potential wells (Thompson & de Souza [51]).

Finally we have the chaotic attractor, which is the simplest manifestation of chaos. This technical expression is used to describe irregular non-periodic motions of a deterministic system with no noise excitation. Chaos is not present in linear systems, but arises typically in nonlinear systems when the dimension of the phase space exceeds two. It is characterised by a noisy broad-band power spectrum and exponential divergence from adjacent initial conditions (the butterfly effect). Despite its random features, the motion of a chaotic attractor lies within a well-defined fractal form, with finer and finer structure revealed under unending magnifications. For a useful introduction to fractals see Falconer [6]. Chaotic attractors are observed in a pendulum subjected to heavy sinusoidal excitation. The chaotic attractor discovered by Lorenz is described in Thompson & Stewart [52], Chapter 11, while a nice illustration of chaos in a driven oscillator due to Ueda is described in Chapter 1 of the same work.

1.3 Basins of attraction

In the phase space of a dissipative system, any number of attractors may coexist, each within its own separate basin of attraction. The phase space will normally be filled by these basins in a cellular fashion. The basin boundaries are formed by the stable manifolds or *insets* of unstable saddle solutions, which act by repelling across a boundary but attracting within it. Thus a (pathological) motion started on one of these insets would flow to the governing saddle. In driven oscillators the governing saddle is often a chaotic saddle, which gives the basin boundaries an infinitely layered fractal structure. Trajectories from adjacent initial conditions in the fractal zone then experience a chaotic transient, before being dispersed to one of two or more coexisting attractors. In engineering resonance problems, such fractal boundaries often have a greater influence on the overall global dynamics than the chaotic attractors, as we shall describe later.

1.4 Local and global bifurcations

A bifurcation of a system is a qualitative change in the phase portrait that occurs at a certain critical value of a control parameter. Local bifurcations are associated with the folding and branching instabilities of attractor paths. Global bifurcations are often associated with changes in the basin structure. Regular bifurcations are well classified and understood, but chaotic bifurcations governing the instability of chaotic attractors and the explosion of fractal basins are a current research topic. These chaotic bifurcations add a further element of unpredictability to nonlinear dissipative systems and have direct relevance to engineering. A recent classification of the generic bifurcations of dissipative dynamical systems is given in Thompson, Stewart & Ueda [54]. This article highlights the ideas of safe, explosive and dangerous bifurcations, which we now outline.

1.5 Safe, explosive and dangerous bifurcations

The so called safe bifurcations are subtle events, with the continuous supercritical growth of a new attractor path. They give no dynamic jump or abrupt enlargement and are determinate, with a single outcome under noise excitation. The basin boundary is remote from the bifurcating attractors and there is no hysteresis, with paths retraced on reversal of the control sweep. They include supercritical forms of the Hopf, Neimark (secondary Hopf) and flip (period-doubling) bifurcations as well as band merging of chaotic attractors.

The explosive bifurcations are discontinuous global events with an abrupt enlargement of the attracting set, but no jump to a disconnected attractor. They have a remote basin boundary and are determinate with no hysteresis. They arise in flows and maps as omega and intermittency explosions and include the regular-saddle and chaotic-saddle explosions encountered in driven oscillators.

Finally the dangerous bifurcations are discontinuous, with the blue-sky disappearance of the attractor. There is a fast dynamic jump to a distant unrelated attractor of any type and hysteresis on reversing the controls. Either the basin shrinks to zero, or the attractor hits the boundary of a residual basin. Dangerous bifurcations can be either determinate or indeterminate, depending on the specific global topology, and we discuss them in more detail in the following section.

1.6   Indeterminate bifurcations

The most unpredictable response of a nonlinear dynamical system is one that starts nominally on a basin boundary. Noise and perturbations can then send trajectories to alternative attractors with totally different characteristics where one might be safe, while another might represent the failure of an engineering system. Now such a start might be thought unlikely, but under the slow sweep of a control there are generic bifurcations that bring a system precisely onto a basin boundary. These are the indeterminate dangerous events that give an unpredictable dynamic jump to one of two or more alternative coexisting attractors.

In a 2D flow, the static fold (saddle node of fixed points) is generically determinate because its 1D outset would require a nongeneric saddle connection to make its outcome indeterminate (Fig. 1). By contrast, the cyclic fold (saddle node of cycles) has a 2D outset, which allows generic connections with any number of saddles giving the possibility of an indeterminate jump to any number of coexisting attractors (Fig. 2). An analytical example of such a generic indeterminacy is given in Thompson [41]. In driven oscillators, cyclic folds can occur on fractal basin boundaries, giving the tangled saddle node that we examine in Section 6.1. The 2D spiral outset of a subcritical Hopf bifurcation also gives generic indeterminacy, and an example in aeroelastic galloping is given in Thompson [41]. The subcritical Neimark (secondary Hopf) bifurcation can likewise be generically indeterminate. An example of an indeterminate subcritical flip has recently been presented by Soliman (to be published).

Similar indeterminacies can also arise in the global dangerous bifurcations. When a stable cycle in a 2D flow collides with a saddle fixed point in a saddle

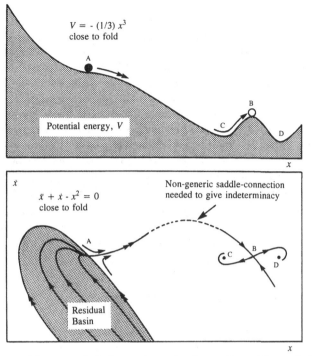

Figure 1: Non-generic indeterminacy in the static saddle node fold.

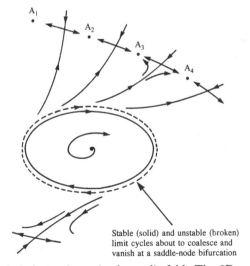

Figure 2: Generic indeterminacy in the cyclic fold. The 2D outflow from the saddle node fold allows any number of generic saddle connections. The jump from the bifurcation can lead to any of the available attractors.

connection, the 1D outflow from the saddle (see Fig. 13.5 of Thompson & Stewart [52] and Fig. 16 of this article) allows only a nongeneric indeterminacy, as with the static fold. However, when a chaotic attractor expands to hit a saddle cycle whose smooth, untangled inset forms its basin boundary we have the regular saddle catastrophe (regular boundary crisis) for which indeterminacy is generically possible. Finally, when a chaotic attractor expands to hit the accessible orbit within a fractal basin boundary we have the chaotic saddle catastrophe (chaotic boundary crisis). Here, generic indeterminacy with a fractal accumulation has been observed in the twin-well Duffing oscillator (Stewart & Ueda [39]), and in the escape equation (see Section 6.1).

Outside this list of codimension-one events are the special bifurcations generated by parametric excitation discussed in Section 6.5.

To guard against the dangers inherent in these indeterminate bifurcations engineers and applied scientists must clearly devote more time and effort into exploring the full range of response of their analytical, numerical and experimental models.

2   A Class of Driven Oscillators

Recent progress in unravelling the complex behaviour of driven nonlinear oscillators derives from the organising role of the invariant manifolds and their global bifurcations. These manifolds, the insets and outsets of unstable saddles, structure the phase space and control the attractors, their basins and their bifurcations. We present these ideas here in the context of a simple archetypal oscillator governing the escape from a potential well.

Many problems in the mathematical sciences correspond to the escape from a potential well. In particular, engineering failures are often triggered when an underlying dynamical system escapes from a well. Electrical systems can slip out of synchronisation. With power generators this can result in the blackout of an entire city; in the phase-locked-loop of a receiver a slip from the locked configuration results in a loss of communication; if a synchronous motor slips under excessive load, time-keeping will be lost. In civil engineering, slender shell structures can buckle elastically under compressive loading, resulting in collapse. In naval engineering, ships can capsize under the lateral wave excitation produced by beam seas.

We shall focus here on the canonical escape equation (Thompson [40]) displayed in Fig. 3 and we shall be interested in its response at different values of the forcing magnitude F and the forcing frequency ω. For most of our present discussion (and unless indicated to the contrary) we shall hold the damping level fixed at $\beta = 0.1$. A comprehensive and detailed examination of the different bifurcation structures at different damping levels for the escape equation (with a cubic potential) and for an oscillator with a twin-well potential is given in Stewart, Thompson, Ueda & Lansbury [38].

All oscillators in an asymmetric well, $V(x) = \int f(x)\,dx$, with the general form

$$\ddot{x} + g(x, \dot{x}) + p_1(t)\,f(x) = p_2(t) \qquad p_{1,2}(t) = p_{1,2}(t+T)$$

exhibit universal features illustrated by the canonical escape equation:

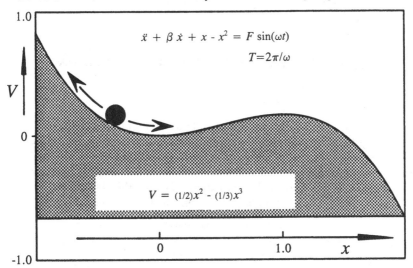

$$\ddot{x} + \beta\,\dot{x} + x - x^2 = F\sin(\omega t)$$
$$T = 2\pi/\omega$$

$$V = (1/2)x^2 - (1/3)x^3$$

Behaviour of autonomous system with $F=0$, $\beta=0.1$:—

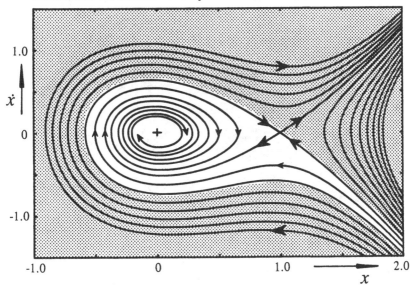

Figure 3: Escape from a potential well, showing the canonical escape equation and its autonomous response.

It should be emphasised that the phenomena that we shall describe are typical of a wide class of systems. This class includes, for example, nonlinear damping, different well shapes and different direct and parametric forcing functions (time-periodic, but not necessarily sinusoidal). Many of the features also arise in hardening systems where there is no escape mechanism at all (Soliman & Thompson [36]).

3 Steady States and Local Bifurcations

3.1 Phase portraits

The lower diagram of Fig. 3 shows the 2D phase portrait spanned by x and y ($= dx/dt$) for the autonomous system with no external forcing ($F = 0$). Motions that spiral in to the point attractor at the bottom of the well are separated from motions that escape to infinity by the inset of the unstable hill-top saddle solution. With light external forcing, the 3D phase space has the form of Fig. 4. The stable equilibrium in the well has been replaced

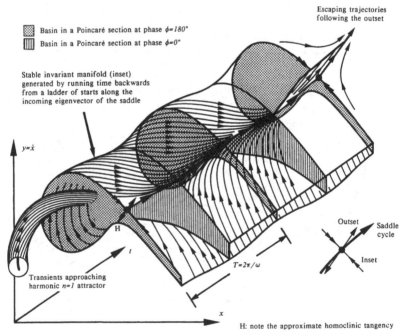

INVARIANT MANIFOLD OF A SADDLE CYCLE FORMING A BASIN SEPARATRIX

Escape equation: $\beta=0.1$, $\omega=1.0$, $F=0.074$ (partly schematic)

Escaping trajectories following the outset

Basin in a Poincaré section at phase $\phi=180°$
Basin in a Poincaré section at phase $\phi=0°$

Stable invariant manifold (inset) generated by running time backwards from a ladder of starts along the incoming eigenvector of the saddle

$y=\dot{x}$

H

Outset — Saddle cycle

Inset

$T=2\pi/\omega$

Transients approaching harmonic $n=1$ attractor

x

H: note the approximate homoclinic tangency

Figure 4: Sketch of the 3D phase portrait of the escape equation, showing the invariant manifold of the hill-top saddle cycle.

by a periodic attractor and the unstable hill-top equilibrium state has been replaced by an unstable saddle cycle. The inset of this saddle is a sheet of solutions forming the boundary of the safe non-escaping basin. This figure also illustrates how we can sample the flow stroboscopically at the forcing period to obtain a Poincaré section and an associated 2D mapping. For a more general illustration, see Thompson & Stewart [52], Fig. 5.1.

3.2   Resonance response diagrams

In engineering, the resonance of a system is often displayed by plotting the maximum displacement as a function of the forcing frequency for given fixed values of the forcing magnitude. This is done for the present system in Fig. 5. At $F = 0.056$ for example, we see the typical hysteresis response of a softening nonlinear oscillator. There is a jump to resonance at fold A as the forcing frequency is increased and a jump from resonance at fold B as the frequency is decreased. In the hysteresis zone between these two cyclic folds there are three steady state periodic solutions with the same frequency as the forcing (referred to as $n = 1$ harmonic solutions): the non-resonant attractor is denoted by S_n, the resonant attractor by S_r and the unstable resonant saddle by D_r.

The behaviour at two higher values of F is shown in Fig. 6. Notice that at these values of F the maximum steady state response amplitude is quite close to the hill-top value of $x = 1$. In the top diagram, two opposing supercritical flip bifurcations generate a closed loop of $n = 2$ subharmonic attractors. In the lower diagram this feature has developed into two opposing cascades to chaos. Such period-doubling cascades are the most common route to chaos and are nicely illustrated by the logistic map (see for example Thompson & Stewart [52], Chapter 9). A summary of the steady state behaviour at three high values of F is finally given in Fig. 7.

In the bottom diagram we observe a frequency interval with no main attractor (though it might contain high-order subharmonics with very small basins of attraction, existing over very small frequency ranges). Here, as we slowly increase the driving frequency, we come to fold A (with a relatively low response amplitude) from which the system would jump straight out of the well. If, conversely, the frequency is slowly decreased from an initially high value, then a period doubling cascade to chaos occurs, followed by a chaotic crisis at which the system again jumps out of the well.

In the top diagram (at lower F), the jump to resonance at fold A always restabilises on the stable resonant state R. Of more interest is the intermediate case displayed in the central diagram. Here the jump from fold A may or may not restabilise on R, the bifurcation being indeterminate with an outcome that depends sensitively on the way it is realised. We shall explain later how at point A the system is sitting precisely on a fractal basin boundary.

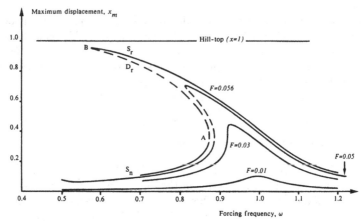

Figure 5: A nest of resonance response curves for the escape equation at relatively low forcing levels.

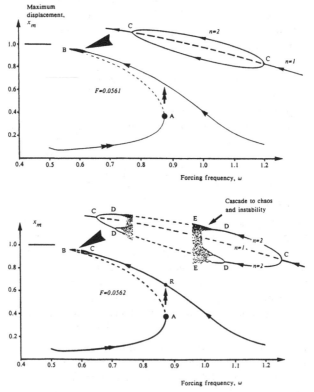

Figure 6: A pair of resonance response curves for the escape equation at two intermediate forcing levels.

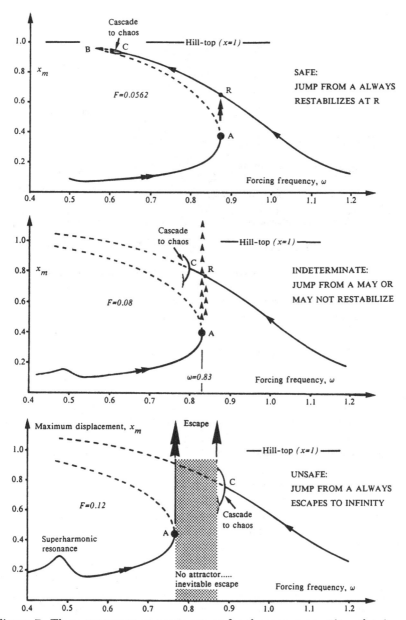

Figure 7: Three resonance response curves for the escape equation, showing safe, indeterminate and unsafe jumps to resonance from the saddle node fold A. (The idea of a safe jump should not be confused with the description of bifurcations as safe, explosive or dangerous: in the bifurcational context, fold A is always classified as a dangerous bifurcation).

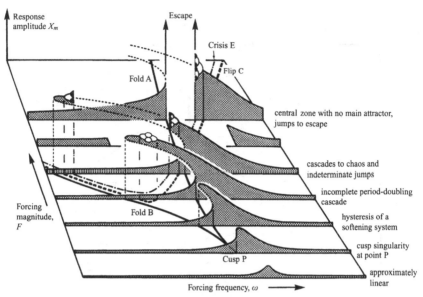

Figure 8: Sketch of the resonance response surface summarising the behaviour of the escape equation, showing bifurcation arcs in the control plane.

3.3   Bifurcation arcs in control space

The complete set of resonance response diagrams can be sketched as a surface, as in Fig. 8, where the bifurcation sets have been projected into the base plane. Fig. 9 shows more details of this (F, ω) control plane. Here the bifurcation arcs summarise the jump to resonance at fold A, the jump from resonance at fold B, the first period-double to a stable subharmonic of order $n = 2$ at flip C, and the final loss of stability of the chaotic attractor at crisis E. Notice how the fold arcs A and B are generated by the cusp at P. On the crisis arc the chaotic attractor is in collision with an unstable subharmonic of order $n = 6$, the remnant of a short-lived fold-cascade-crisis scenario. This arc can be approximately located by numerically following the Birkhoff signature change (see Section 5.5) that occurs on arc S. The intersection of arcs A and E at Q marks the point of optimal steady-state escape.

If F is greater than F^Q the resonance response curve will have a frequency interval with no (main sequence) attractor as in the lower diagram of Fig. 7. Here under increasing ω the jump to resonance from fold A will inevitably carry the system straight out of the well, as will the jump from crisis E under decreasing frequency. At a value of F somewhat less than F^Q we have indeterminate bifurcations, and the jumps may or may not restabilise onto the available attractors (middle diagram of Fig. 7). To understand this indeterminacy we shall need the perspective supplied by the global organisation of phase space that we shall address shortly. Further useful information about

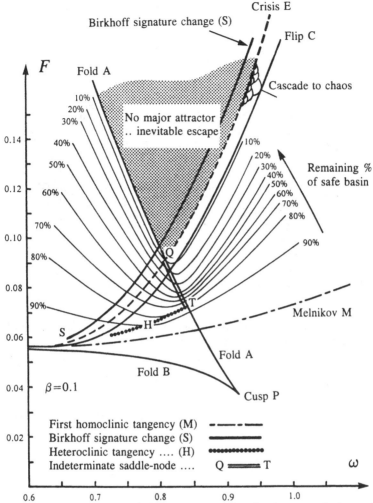

Figure 9: Bifurcation arcs of the escape equation in the control plane. Superimposed contours show the remaining percentage of the safe basin.

the bifurcational structure of driven oscillators can be derived from the theory of knots and braids (Ghrist & Holmes [8], McRobie [19], McRobie & Thompson [22, 23]). Modern comprehensive articles on the bifurcations of the escape process are Stewart, Thompson, Ueda & Lansbury [38], Thompson & McRobie [45], Ueda, Yoshida, Stewart & Thompson [56].

3.4 Crisis of the chaotic attractor

A bifurcation that warrants further comment is crisis E, at which the chaotic attractor created by a period-doubling cascade finally loses its stability. The

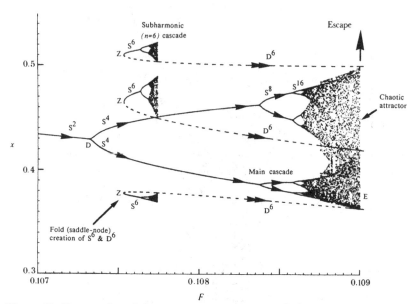

Figure 10: Destruction of the main sequence chaotic attractor at the boundary crisis E, by collision with an unstable subharmonic of order 6. Notice that the mapping coordinate is here sampled at twice the forcing period.

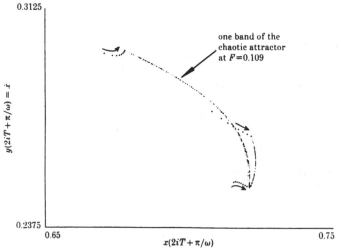

Figure 11: Boundary crisis E in a section sampled at twice the forcing period. Arrows show the movement of 3 mapping points of the subharmonic saddle as the forcing magnitude is stepped up to 0.109. The chaotic attractor does not move appreciably during this stepping.

mechanism of this crisis, typical in driven oscillators, is illustrated in Fig.
10. A stable-unstable pair of subharmonic orbits, here of order 6, is created
out-of-the-blue by a cyclic fold (notice that in the coordinate range of Figs.
10 and 11, only 3 of the 6 Poincaré mapping points are visible). This type
of event happens all the time in heavily driven nonlinear oscillators, but the
events are not easily found numerically because the stable orbits often have
very small basins, and only exist in a very small parameter regime. They
can appropriately be called fugitive subharmonics. The stable $n = 6$ orbit
period doubles to a six-band chaotic attractor, which loses its stability at a
crisis when it hits the unstable $n = 6$ orbit, denoted by D^6. Long after this
fold-cascade-chaos-crisis of the $n = 6$ solutions, it is the remaining D^6 which
eventually collides with and destroys the main chaotic attractor (Thompson
& Ueda [55]). A Poincaré section near the final collision is shown in Fig. 11.

4 Transients and Basins

4.1 Basin erosion phenomenon

Engineers and applied scientists must pay attention, not only to the steady
states, but also to the transient motions that can arise from finite disturbances, sudden changes in excitation, etc. In fact a global view of all possible
transient motions supplies a view of the dynamics that is both more robust
and more relevant than a detailed inspection of the steady states and their
local bifurcations.

In examining transients, it will suffice here to focus on the long-term behaviour asking the basic question: will the system escape or not from a given
initial condition if we were to wait for an infinite time? The more practical
question concerning escape in a finite number (m) of forcing cycles can often
be fairly accurately assessed once this basic question about the basins of attraction has been answered. Conversely, we have found that when studying
the overall morphology of the safe basin, the escaping set for $m = \infty$ is approximated with good scientific accuracy by studies with m in the interval
8 to 64. Notice however that while an $m = \infty$ safe basin can experience an
instantaneous loss of area under increasing F (see Section 6.3), this will not
be observed in finite m simulations.

We turn then from the steady state behaviour of our escape equation to
the transient behaviour as represented by the safe basin of attraction. This
basin is dramatically eroded by incursive fractal fingers, as illustrated in Fig.
12. Here safe basins of attraction at increasing levels of F have been determined numerically using a grid of $[x(0), y(0)]$ values. Initial conditions
leading to escape over the hill-top are denoted by white, while initial conditions which settle onto finite motions within the well (harmonic, subharmonic
or chaotic) are denoted by black. We observe that as F is increased up to

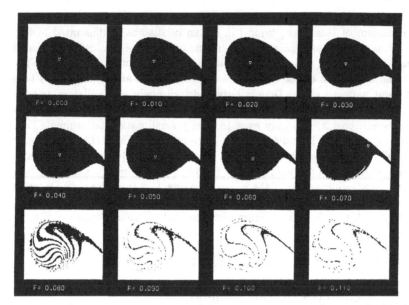

Figure 12: Basin erosion sequence of the escape equation, showing the black
safe basin in the space of the initial conditions $[x(0), y(0)]$ as the forcing
magnitude is increased.

0.07, only a small reduction in the safe region occurs. But at about this
value, a homoclinic tangency is formed (as described in Section 5.1), creating
fractal fingers which soon sweep dramatically across the safe basin so that at
$F = 0.08$ most of the safe basin has been destroyed. This process is nicely
shown in a computer-generated video made by Cusumano, who has also made
ingenious experimental observations of the associated phenomena (Cusumano
& Kimble [5]). Details of the fractal incursion are shown in Fig. 13 which was
produced, not by using a grid of initial conditions, but by locating the fractal
boundary defined by the inset of the hill-top saddle cycle. This was done
numerically by running time backwards from a ladder of initial conditions
on the appropriate eigenvector of the saddle. A characteristic feature of the
(highly discontinuous) erosion process is the fold-cascade-crisis appearance of
subharmonics at the tips of the fingers. An $n = 3$ subharmonic is seen in
the last picture of Fig. 13 which also shows the Poincaré mapping sequence
$0, 1, 2, \ldots, 7$ by which a trajectory steps from finger to finger before escaping
to infinity.

4.2   Engineering integrity curves

The simplest way to quantify this basin erosion is to evaluate the area of the
safe basin. This must be done within a suitable $[x(0), y(0)]$ window, because

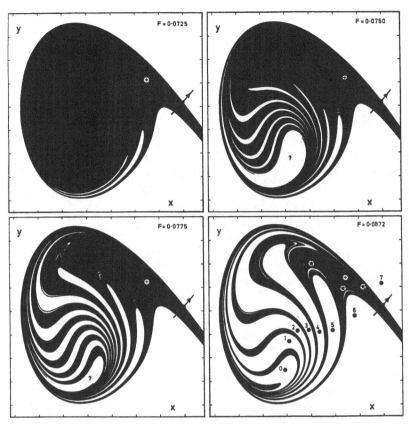

Figure 13: Development of the homoclinic tangle, showing how incursive fractals sweep across the centre of the safe basin. The cross denotes the main-sequence harmonic (period one) attractor. The stars denote a subharmonic attractor of period 3. A typical mapping sequence to escape is denoted by points $0, 1, 2, \ldots, 7$.

the safe basin has infinite area (due to the positive divergence of a damped system under time reversal). The area is one measure of the integrity of the system against disturbances (others are described in Soliman & Thompson [31]), and a plot of the area against F can be called an integrity diagram. In Fig. 14 the basin area (normalised with respect to the autonomous system) is plotted against the forcing magnitude. Up to the homoclinic tangency we have a smooth boundary and there is little reduction in basin area. Shortly after the tangency we observe the rapid erosion process giving us the *Dover cliff* fall in basin area. This cliff corresponds to a well-defined loss of integrity and can be used as a design criterion. We would not want to operate an engineering system at values of F greater than that of the cliff face, even though there are still stable periodic motions within the well. The Dover

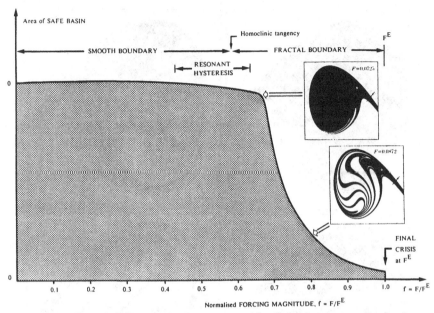

Figure 14: Dover cliff in the integrity curve of safe basin area versus forcing magnitude. Superimposed sections show the fractal incursion process.

cliff is observed for a wide class of driven oscillators, and single and twin well potentials exhibit remarkably similar cliff profiles (Lansbury & Thompson [14]). Since the fractal fingers sweep rapidly across the centre of the section, the forcing magnitude of the cliff can be approximately located by running just one time simulation from the ambient state $[x(0), y(0)]$, rather than a complete grid. This gives a quick and economical way of locating the loss of engineering integrity. It has been proposed, for example, as a useful tool in the simulation and laboratory testing of model ships for capsize due to surface waves (Thompson, Rainey & Soliman [46, 47], Rainey & Thompson [28], MacMaster & Thompson [16]).

4.3   Fractals in control space

The trajectories from the ambient state that we have just been discussing have a deeper dynamical significance, because they define a fractal basin boundary in control space. At a fixed damping level, the safe basin of the escape equation can be thought of as a master basin (Thompson & Soliman [49]) in the 4D phase-control space spanned by $x(0)$, $y(0)$, F and ω. The phase space basins of Figs. 12 and 13 are cross-sections of this master basin corresponding to fixed values of F and ω. The control space basins that we are now considering are likewise just cross-sections of the master basin, this time at the fixed values of $x(0) = 0$, $y(0) = 0$.

The left hand picture of Fig. 15 shows this control space cross-section of the safe basin of the escape equation: here safe is denoted by white, unsafe by black. Also shown on the figure are the steady state bifurcation arcs. Notice that thin black fingers extend down to point T, which is the lower limit for indeterminate jumps to resonance governed by a heteroclinic connection between the hill-top saddle and the resonant saddle. We shall say more about this in a later section and further details can be found in Soliman [30], from which this diagram has been reproduced. An experimental study of this control-space cross-section is reported by Gottwald, Virgin & Dowell [10]. To emphasise just how universal these features are, we show in the right hand diagram an equivalent picture for a parametrically excited system, also from Soliman [30].

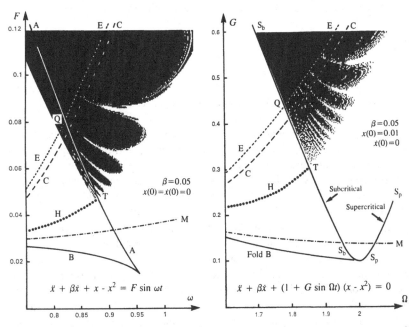

Figure 15: Fractal boundaries in control space for the escape equation and for a similar parametrically excited system. These pictures are based on the work of Soliman [30]. Notice that the damping level for the escape equation is here 0.05.

5   Global Bifurcations

5.1   The homoclinic tangency

In a 2D flow, an outset of a saddle can return as an inset of the same saddle, as illustrated in Fig. 16. In this example, the saddle collides with an expanding stable cycle, which vanishes into-the-blue at the global bifurcation. The equivalent event in a 2D mapping is the homoclinic tangency, illustrated in Fig. 17, which might represent three stroboscopic Poincaré sections of the escape equation under increasing F. In such a section the flow of the previous figure is replaced by a differentiable mapping (a diffeomorphism), and by considering the forward and backward iterations of the map it is clear that if the inset touches (or intersects) the outset once, it will touch (or intersect) it an infinite number of times. This fact, together with the asymptotically slow dynamics near the saddle D implies that after the tangency there will be an infinite number of intersection points accumulating onto D, as shown. After a homoclinic tangency we say that we have a homoclinic tangle. The 3D flow manifolds of such a tangle are shown schematically in Fig. 18.

5.2   Homoclinic tangles and fractal boundaries

Now for the escape equation we have seen that at $F = 0$ the basin of attraction of the stable equilibrium at the bottom of the well is bounded by the trajectory that flows asymptotically into the unstable hill-top equilibrium. As F is increased from zero, this hill-top equilibrium is transformed into a small unstable oscillation across the hill-top, denoted by D_h, and its inset continues to delineate the boundary between escaping and non-escaping initial conditions. Under further increase of F, this boundary undergoes a series of metamorphoses at global bifurcations, the most significant of which are the homoclinic tangency M and the heteroclinic tangency H, which occur on the arcs shown in Figs. 9 and 15 in the (F, ω) control space.

At the homoclinic tangency M, the inset and outset of the hill-top saddle cycle D_h first touch one another. For $F < F^M$ any small inwards disturbance of D_h gives a trajectory that falls onto one of the safe, bounded attractors within the well as illustrated in the top section of Fig. 19. At the homoclinic tangency of the second section, the forcing is just strong enough to allow one (disturbed) trajectory to depart asymptotically from D_h along an outgoing eigenvector, pass across the floor of the well, and return asymptotically to D_h along an incoming eigenvector. At the higher F of the third section, some initial conditions from the outgoing eigenvector will be safe, while others will give escape.

Throughout this sequence of events, the inset of D_h (strictly its closure) remains the boundary of the safe basin, but beyond M the homoclinic tangling gives it a fractal nature (McDonald, Grebogi, Ott & Yorke [17]). An infinite

$$\ddot{x} + \beta\dot{x} + \gamma\dot{x}^3 + x - x^2 = 0$$

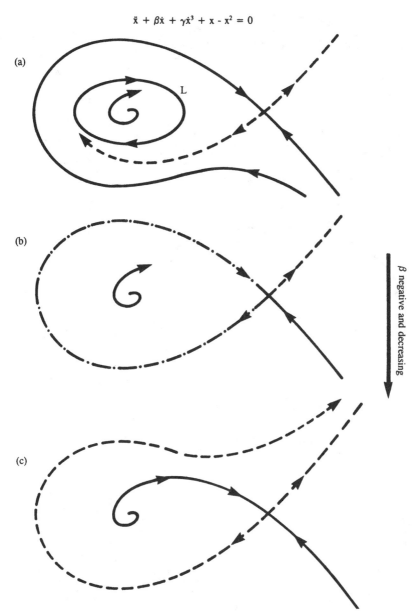

(a) As β is decreased through zero, positive γ gives a supercritical Hopf bifurcation generating the growing stable limit cycle, L. (b) On further decrease of β, cycle L collides with the saddle in a saddle connection. (c) After the saddle connection there is no attractor.

Figure 16: Illustration of a generic saddle connection in the 2D flow of an autonomous oscillator.

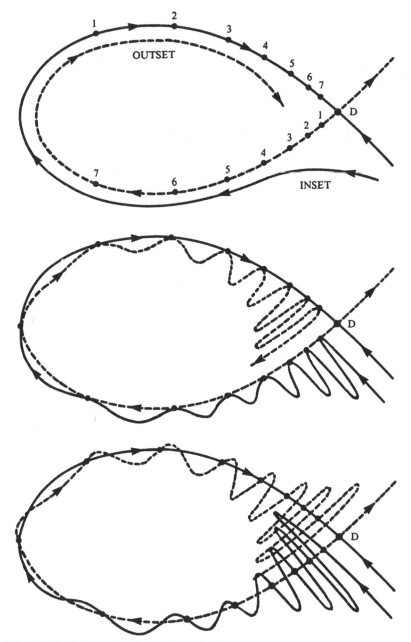

Figure 17: Schematic illustration of a homoclinic tangency in a Poincaré section. Here D is a directly unstable saddle cycle, and invariant manifolds are sketched before, at and after the tangency.

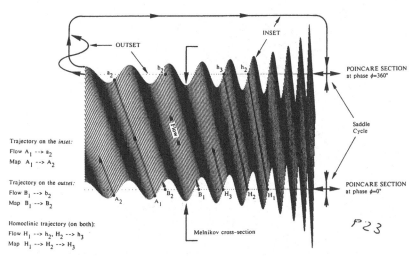

Figure 18: Schematic illustration of the invariant manifolds in a homoclinic tangle, showing the flow between two Poincaré sections.

number of infinitely thin tails of the escaping and non-escaping basins are intertwined around the edge of the safe basin, making prediction impossible within this fractal fringe. The structure of the invariant manifolds within a homoclinic tangle is amenable to analysis by the techniques of lobe dynamics (McRobie & Thompson [20, 21]). The powerful contributions of Allan McRobie show how the manifolds of the 3-striped Smale horseshoe can be used to locate orbits and establish important bifurcational precedences (McRobie [19], McRobie & Thompson [23]). These methods have recently been applied to the problem of the parametrically excited pendulum in Bishop & Clifford [3].

5.3   Fractal boundaries around regular states

We have seen that at the first homoclinic tangency at F^M, the union of the basins of the non-escaping attractors acquires a fractal boundary. So for F marginally above F^M, there will be a thin fractal fringe around the safe non-escaping basin. We should emphasise, however, that this fringe has initially (before the subsequent incursion) very little effect on the overall dynamics of the oscillator. At the centre of the safe basin there will often still be a regular period-one attractor whose local transients are entirely normal. The period-doubling to chaos of the central main-sequence attractor typically occurs long after the first tangency ($F^C \gg F^M$) and is a totally unrelated event.

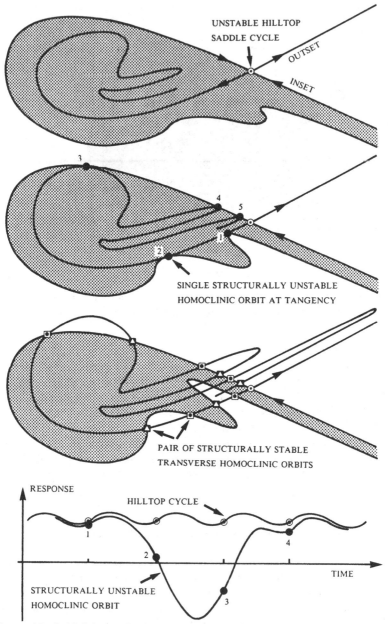

Figure 19: Initial homoclinic tangency of the hill-top saddle cycle. Three Poincaré sections before, at and after the tangency show how the safe basin acquires a fractal boundary. The lower picture shows the time history of the homoclinic orbit.

This being the case, we must ask what meaning can be attributed to the often heard, and usually misunderstood statement, that a homoclinic tangency implies chaos? The answer is that associated with the Smale horseshoe created at the tangency there will be chaotic transients and highly localised fold-cascade-crisis scenarios, all however located within the thin fractal zone around the edge of the basin.

5.4 Melnikov perturbation analysis

The first homoclinic tangency at F^M can be estimated analytically using the Melnikov perturbation method. This can be interpreted as an energy balance approach (see Appendix A), in which the energy input due to the forcing is equated to the energy lost due to dissipation. These energies are calculated by integrating the work done as the real system is imagined to move around the saddle-loop of the unforced and undamped Hamiltonian system. Applied to the escape equation, the Melnikov arc in the (F, ω) control space at the damping level of $\beta = 0.1$ is almost coincident with the true arc of homoclinic tangency over the frequency regime of Fig. 9. At higher ω the prediction becomes less accurate (see Foale & Thompson [7], Fig. 1).

5.5 Birkhoff signature change

Under further increase of F, the inset and outset of the hill-top saddle D_h become increasingly tangled. Further secondary homoclinic tangencies arise, a key tangency being that at which there is a change in the period-one Birkhoff signature (McRobie [18]). This is illustrated in Fig. 20 and loosely corresponds to the lowest F at which a trajectory can depart from the outgoing eigenvector of D_h, make one complete oscillation within the well during one period of the applied forcing, and return to the incoming eigenvector of D_h. This Birkhoff signature change occurs on the global bifurcation arc S of Fig. 9, lying just above the crisis arc E. Thus arc S can be used as a guide to the location of arc E, as discussed in McRobie [18]. A similar situation has subsequently been identified in the parametrically excited pendulum (Clifford & Bishop [4]).

5.6 Heteroclinic tangencies and heteroclinic chains

We have seen that in a 2D flow a saddle can be connected to itself by a homoclinic loop. In such a flow, two distinct saddles can also be linked by a heteroclinic connection, as in Fig. 21. This can give an instantaneous change to the basin structure, as illustrated. An illustration of such an event in an undriven oscillator is shown in Fig. 22.

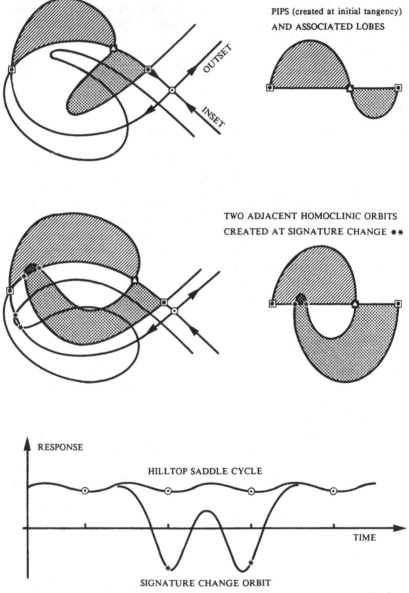

PIPS (created at initial tangency)
AND ASSOCIATED LOBES

TWO ADJACENT HOMOCLINIC ORBITS
CREATED AT SIGNATURE CHANGE ✳✳

Figure 20: Schematic illustration of two Poincaré sections showing the first period-one Birkhoff signature change. At the top is shown the basic signature after the initial homoclinic tangency. In the middle is shown the new signature after the internal homoclinic tangency. A time history of the signature-change orbit is shown in the lower diagram.

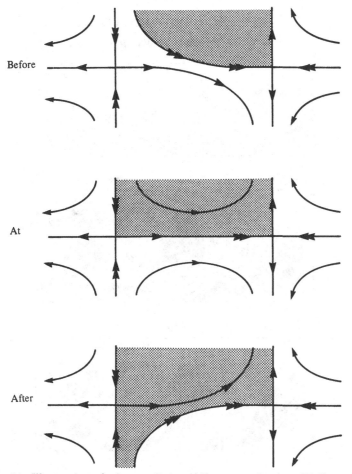

Figure 21: Illustration of a heteroclinic saddle connection in a 2D flow, showing the corresponding basin metamorphosis.

The corresponding event in a 2D mapping is the heteroclinic tangency illustrated in Fig. 23. Like the homoclinic tangency, this has an infinite number of tangencies (or intersections) that accumulate onto the saddles. If the heteroclinically tangled inset forms a basin boundary, an infinite number of infinitely thin fingers of the basin will accumulate onto the saddle as illustrated. Notice however that this is a smooth accumulation (sometimes called a mosquito coil) which does not of itself generate any fractal structure. Heteroclinic connections in driven oscillators often form chains, as illustrated in the lower picture of the figure. Notice that in this chain, a small perturbation of the lower-left saddle can give a trajectory that ends up close to the upper-right saddle.

The Pendulum with Torque
$$\ddot{x} + \beta\dot{x} + \sin x = \mu, \quad \mu = 0.24$$

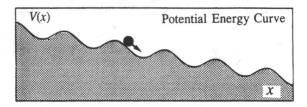

Light damping, $\beta = 0.1$ ➤➤ Steady Motion

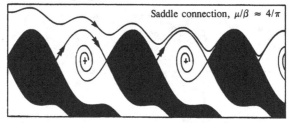

Critical damping, $\beta = 0.19$

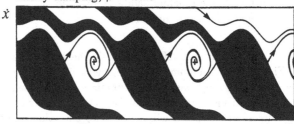

Heavy damping, $\beta = 0.25$

Figure 22: An example of a heteroclinic saddle connection for a ball rolling on a corrugated surface. This is the same as a pendulum subjected to an applied torque, though for the pendulum it could be regarded as a homoclinic connection.

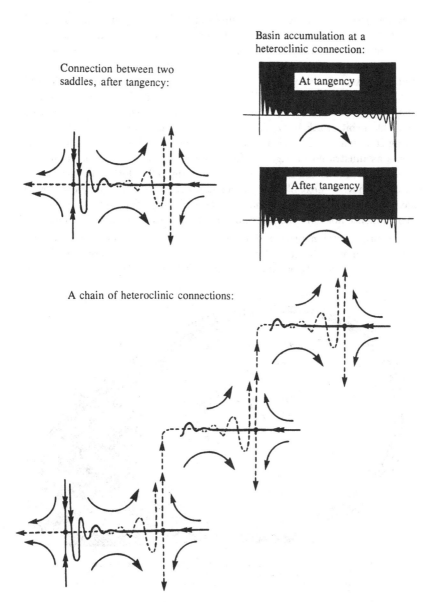

Figure 23: Illustration of heteroclinic connections in a 2D map. The lower picture shows how a series of such connections can form a heteroclinic chain.

5.7   Accessible orbits and basin metamorphoses

A major practical concern in contemplating a fractal basin boundary is the
extent of the fractal fringe. Can we be sure that no thin escaping tails penetrate into the centre of the safe basin? The good news is that the penetration
of the tails is always blocked by their accumulation onto an unstable saddle
solution, heteroclinically tangled with D_h, that is called the accessible orbit
(Grebogi, Ott & Yorke [11]). The bad news is that this accessible orbit is
usually very hard to locate numerically. Under increasing F, the role of the
accessible orbit is taken by a sequence of different subharmonic saddles, allowing the tails to make a corresponding sequence of implosions into the safe
regime as illustrated in Fig. 24. Here we see how a heteroclinic connection
from A to B allows the fingers to implode: originally blocked by the inset of
saddle A, they are next blocked by the inset of saddle B. The very bad news is
that to determine the region that is unpermeated by the escaping fingers we
need to locate not only the accessible orbit, but also its entire inset manifold!

The basin erosion process down the Dover cliff involves a chain of such
events, as shown schematically in Fig. 25. This figure is based on the fine
manifold studies of Lansbury, Thompson & Stewart [15]. Just prior to the

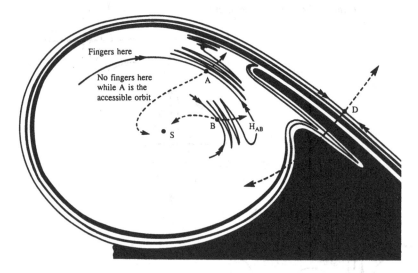

D ... the main governing (hilltop) saddle, whose homoclinic tangle has generated the
black/white fractal basin boundary.

A ... the first accessible saddle orbit, whose inset blocks the penetration of the fractal
fingers.

B ... the second accessible saddle orbit, which takes over the blocking role from A when
the second heteroclinic connection H_{AB} is made.

S ... the central periodic attractor.

Figure 24: A sketch of a Poincaré section showing how the accessible orbits
block the penetration of fractal fingers.

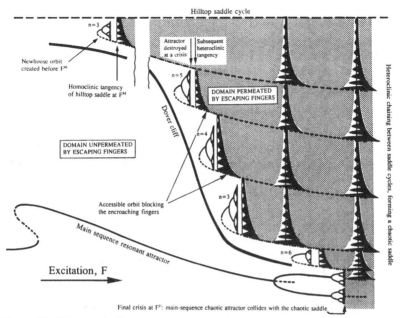

Figure 25: Schematic illustration showing the series of basin implosions associated with discontinuous changes of the accessible orbit during the Dover cliff erosion process. The diagram is based on numerical results for the twin-well oscillator.

homoclinic tangency at F^M where the basin boundary becomes fractal, we see an $n = 3$ fold-cascade-crisis scenario which leaves behind a directly unstable period-three orbit. At the first homoclinic tangency of the hill-top saddle D_h, there is a simultaneous heteroclinic connection between this orbit and D_h which makes it the first accessible orbit. The fractal escaping fingers thus accumulate onto this orbit and are thereby blocked from further encroachment into the safe basin. Subsequent progress down the Dover cliff is characterised by discontinuous changes in the accessible orbit. At the top of the cliff we see for example, an $n = 5$ fold-cascade-crisis scenario, the fold of which creates instantaneously a small basin for the short-lived $n = 5$ subharmonic attractor and its progeny in the outer region of the safe basin. When the subsequent period-five chaotic attractor is destroyed at a crisis, this basin reverts temporarily to the main safe basin, but shortly afterwards there is a heteroclinic connection between the unstable $n = 5$ orbit created at the fold and the current accessible orbit. At this heteroclinic tangency the $n = 5$ subharmonic becomes the new accessible orbit onto which escaping fingers are accumulated and thereby blocked. This mechanism of implosion is repeated down the cliff face as shown. At each stage, the inset of the current accessible orbit blocks the encroaching escape fingers, dividing the safe basin into an outer domain, permeated by (though not completely filled by) escaping fingers, and an inner domain which is unpermeated and completely free from escape fingers. The subharmonics of this process are often located at the tips of the incursive fractal fingers, like the $n = 3$ subharmonic denoted by stars in Fig. 13. This is nicely seen in Lansbury, Thompson & Stewart [15], Fig. 8.

5.8   Role of the chaotic saddle

We should notice here that the heteroclinic chain that exists between the hill-top saddle and the current accessible orbit forms a chaotic saddle. It is the collision with the current accessible orbit, and hence with this chaotic saddle (Stewart [37]) that finally destroys the main-sequence chaotic attractor at the boundary crisis at F^E. This explains why the final main-sequence chaotic oscillation is observed to be remote from the unstable hill-top saddle oscillation at the instant of its final crisis (see for example Thompson [40], Fig. 12). Poincaré sections just before and just after the cliff implosions are given in Lansbury, Thompson & Stewart [15]. The straddle-orbit location of a chaotic saddle is described in Mitsui, Ueda & Thompson [24].

5.9   Heteroclinic connection to the resonant saddle

The basin erosion process that we have just described corresponds to a value of the forcing frequency above that of the cusp P (Fig. 9), where there is no regime of hysteresis. At frequencies below that of P, the existence of the resonant saddle D_r can complicate the story. In particular the heteroclinic chain can reach D_r, giving it a heteroclinic link to the hill-top saddle D_h. The heteroclinic tangency that creates this major heteroclinic connection is shown as arc H in Fig. 9. This arc ends at point T where it hits the line of fold A on which D_r is annihilated at a saddle node bifurcation. Arc H, and its imagined extension beyond the fold line, gives a good estimate of the steepest region of the Dover cliff which can be identified on the figure by the clustering of the contour lines. A heuristic explanation of this is outlined in Soliman & Thompson [33].

This heteroclinic tangency is illustrated in Fig. 26. It corresponds to the first value of F for which a perturbation of D_r can reach D_h. Beyond this value of F escaping fingers accumulate onto the resonant saddle D_r and some outward perturbations of D_r lead to escape, while others still lead to the resonant attractor S_r.

Once we are above the arc H in Fig. 9, D_r is chained to D_h and when D_r is subsequently annihilated at fold A, two important phenomena are observed. On the one hand, there is an instantaneous loss of safe basin area as fractal fingers sweep through the residual basin of the annihilated non-resonant attractor S_n (as we shall see later in Figs. 28 and 29). On the other hand, the jump from fold A, where the non-resonant attractor S_n collides with D_r, is indeterminate, as we shall now describe.

TWO PRECEDING HOMOCLINIC CONNECTIONS OF D_r ★

FINAL (FUNDAMENTAL) HETEROCLINIC TANGENCY ... 1, 2, 3, 4, 5, ...

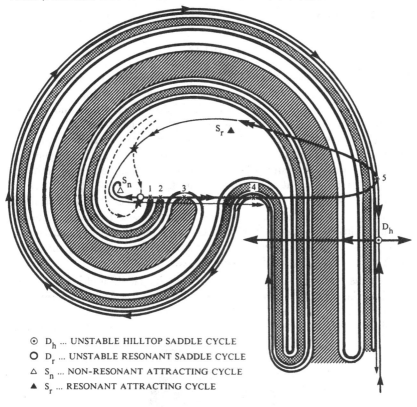

⊙ D_h ... UNSTABLE HILLTOP SADDLE CYCLE
○ D_r ... UNSTABLE RESONANT SADDLE CYCLE
△ S_n ... NON-RESONANT ATTRACTING CYCLE
▲ S_r ... RESONANT ATTRACTING CYCLE

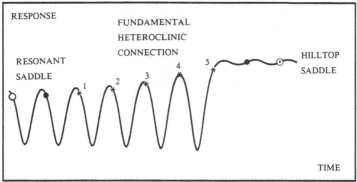

Figure 26: Sketch of the heteroclinic tangency between the resonant saddle and the hill-top saddle. The upper diagram shows the Poincaré section, while the lower diagram shows the corresponding time history.

6   Indeterminate Bifurcations in Driven Oscillators

6.1   Tangled saddle node bifurcation

We are now in a position to explain why the jump to resonance at the cyclic fold A in the middle picture of Fig. 7 is indeterminate. Referring to Fig. 9, we see that this realisation of fold A lies in the interval between Q and T. The invariant manifolds are therefore homoclinically and heteroclinically tangled and at the coalescence of S_n and D_r the system finds itself sitting precisely on a fractal basin boundary as illustrated schematically in Fig. 27. Depending sensitively on how the bifurcation is realised, the jump may settle onto the large amplitude $n = 1$ attractor S_r (denoted by R in Figs. 7, 27 and 30), or may escape out of the well. It may also settle onto a coexisting subharmonic of order $n = 3$ (which is not shown in the resonance response pictures). The manifolds and basins just prior to this event are shown in Thompson & Soliman [50], Fig. 7, and the $x(t)$ time histories of three different jumps from fold A are shown in Fig. 5 of the same paper.

As far as any local analysis is concerned, the saddle node that we are examining here is a perfectly regular one. The high degree of indeterminacy that arises when the system, under a slow sweep of either ω or F to the fold, finds itself sitting on a fractal basin boundary, can only be predicted or

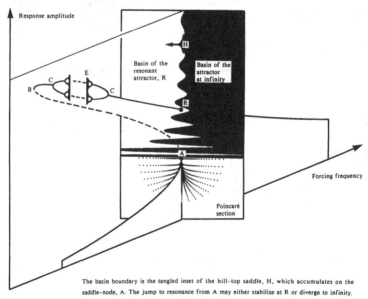

Figure 27: Schematic representation of an indeterminate jump to resonance from a tangled saddle node bifurcation.

understood by an examination, as above, of the global manifold structures. In view of the homoclinic and heteroclinic tangling involved, we have called the event a tangled saddle node bifurcation (Thompson & Soliman [50], Soliman & Thompson [32]).

6.2 Striation of the residual basin

It is instructive to see what happens to the basins at this tangled saddle node bifurcation. The basins just before and just after fold A are shown in Poincaré stroboscopic section in Fig. 28. In the first section we see the node N (representing S_n) close to the saddle S (representing D_r), the latter lying on the boundary of the (large white) residual basin of N. Accumulated onto the far side of S are black and white fractal fingers. Initial conditions in the white fingers settle onto attractors within the well, namely the resonant attractor R (S_r) or the coexisting $n = 3$ subharmonic. Initial conditions in the black fingers escape over the hill-top within the displayed number of forcing cycles. In the second section we are past the bifurcation (having made here a small change of F, rather than varying the frequency) and we can observe the striation of the residual basin. At the bifurcation, broad black and white fingers sweep with infinite speed across the residual basin as we shall now show by reference to a one dimensional mapping.

6.3 Vertical cliff at the tangled saddle node

Close to the saddle node fold, the Poincaré map is essentially one-dimensional, so the mechanism of striation can be understood in terms of the schematic 1D mapping diagrams of Fig. 29. Here it is easy to see that as F decreases towards its critical value, the stripes in the residual basin will acquire infinite speed due to the slow dynamics of the central tunnelling motion. This implies that the fingers of the master basin in phase-control space have finite thickness in phase but are infinitely thin in the control direction. An analysis of this for a cyclic fold in an autonomous 2D flow is give in Thompson [41].

The well-defined black fraction of the residual basin is instantaneously lost from the safe basin of the saddle node. This implies a vertical drop in the integrity diagram of safe basin area versus F at the tangled saddle node bifurcation, as shown. Notice that such an instantaneous drop would only be observed if transient motions were allowed to run for an infinite length of time. Integrity curves plotted on the basis of escape within a finite number of forcing cycles would show a steep but not a vertical drop.

6.4 An indeterminate boundary crisis

In our discussion so far we have focussed on the indeterminacy that exists in the jump to resonance from fold A. An entirely similar indeterminacy arises

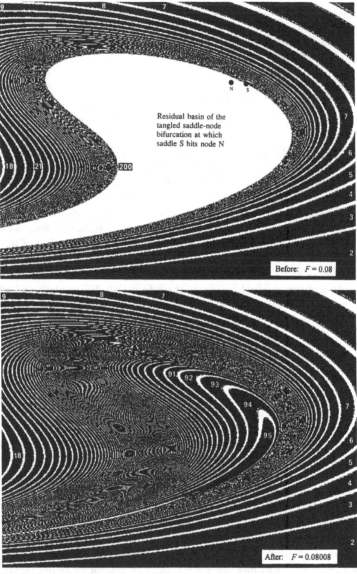

Figure 28: Striation of the residual basin of the tangled saddle node bifurcation. The top picture shows the Poincaré section just before the bifurcation, while the bottom picture shows the section just after the bifurcation. Here black denotes escape within the displayed number of forcing cycles. The main central white area of the top picture is the residual basin of the node N: initial conditions in the thin white stripes settle onto either the resonant harmonic attractor R or the coexisting $n = 3$ subharmonic.

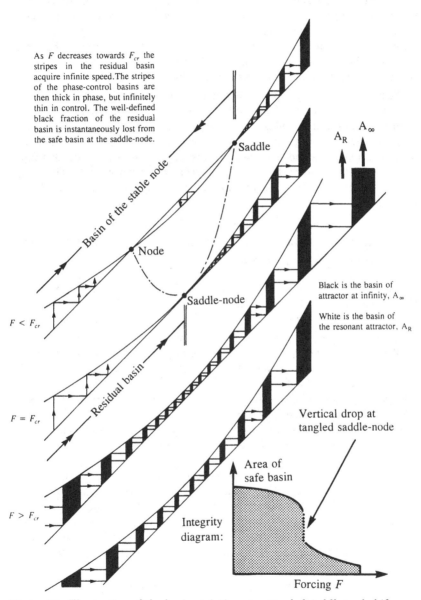

As F decreases towards F_{cr} the stripes in the residual basin acquire infinite speed. The stripes of the phase-control basins are then thick in phase, but infinitely thin in control. The well-defined black fraction of the residual basin is instantaneously lost from the safe basin at the saddle-node.

Saddle

A_R A_∞

Basin of the stable node

Node

Saddle-node

$F < F_{cr}$

Residual basin

Black is the basin of attractor at infinity, A_∞

White is the basin of the resonant attractor, A_R

$F = F_{cr}$

$F > F_{cr}$

Vertical drop at tangled saddle-node

Area of safe basin

Integrity diagram:

Forcing F

Figure 29: Illustration of the basin striation at a tangled saddle node bifurcation using a one-dimensional mapping. A well-defined fraction of the residual basin is lost from the safe basin, giving a vertical drop to an integrity curve.

in the jump from resonance from the crisis E, as illustrated in Fig. 30.

Now as we have shown, the jump from fold A is indeterminate over the interval from Q to T, namely once we are above arc TH where the resonant saddle D_r is chained to the hill-top saddle D_h. In a similar way, the jump from crisis E is indeterminate over the interval from Q to S, once we are above arc SJ where the destroyer governing E is chained to D_h. This destroyer is the colliding unstable subharmonic saddle described in Section 3.4.

Thus the resonance response diagram at the value of $F = 0.08$, shown in the middle picture of Fig. 7 and in the lower picture of Fig. 30, has two indeterminate jumps, at A and E respectively. Approaching E, under decreasing frequency, the system period-doubles to chaos and the jump from resonance at crisis E (a chaotic saddle catastrophe) is indeterminate by virtue of a fractal accumulation (Stewart, Thompson, Ueda & Lansbury [38]). The jump from E either stabilises onto the non-resonant attractor N (namely S_n) or escapes to infinity.

6.5   Indeterminacy in hardening and softening systems under parametric excitation

An extensive survey of parametrically excited systems by Soliman has identified a number of new indeterminate bifurcations. Of these, the subcritical and transcritical forms lie strictly outside the list of codimension-one events, being rather special bifurcations generated by parametric excitation.

The indeterminate subcritical bifurcation arises in parametrically excited oscillations within a cubic well when the driving frequency is approximately twice the linear natural frequency of the system (Soliman & Thompson [34]). This is the regime of the main (principal) parametric resonance that is of greatest interest to engineers. Bifurcations in control space are shown in the right-hand picture of Fig. 15, where the governing equation is displayed. Here we see a rather similar sequence of events to that of our escape equation shown in the left-hand picture. The regime of the indeterminate subcritical bifurcation extends from Q to T, where T is the intersection with an arc of heteroclinic tangency. Notice that the fractal fingers of transient escape extend down to T in both diagrams. This gives us a quick way to locate T, as discussed in Soliman [30]. Indeterminate subcritical bifurcations have also been found in parametrically excited hardening systems with a restoring force equal to $x + x^3$ (Soliman & Thompson [36]). Here there is no hill-top in the continuously hardening potential function.

Transcritical indeterminacies (Soliman & Thompson [35]) have been identified in softening oscillators (governed again by the equation displayed in the right-hand picture of Fig. 15) in the smaller resonance observed when the driving frequency is approximately equal to the linear natural frequency of the system. An example of an indeterminate subcritical flip has recently

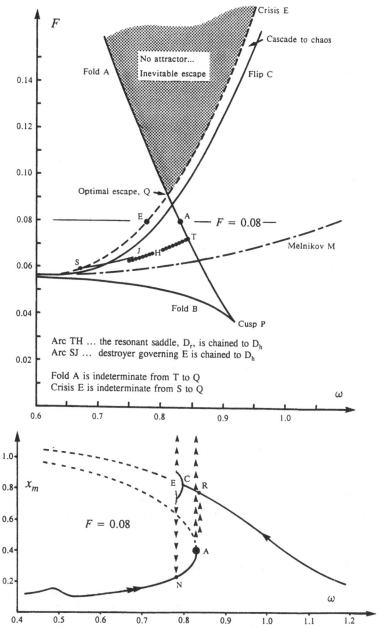

Figure 30: Indeterminate jumps to and from resonance in the escape equation. The indeterminacy at the crisis E is described in Stewart, Thompson, Ueda & Lansbury [38].

been observed by Soliman (to be published), for an oscillator in a cubic well subjected simultaneously to direct and parametric excitation.

A   Melnikov Analysis as an Energy Balance

We consider the escape equation

$$\ddot{x} + \beta\dot{x} + x - x^2 = F\sin\omega t \qquad (A.1)$$

and we define $y \equiv \dot{x}$. We want to estimate the first homoclinic tangency between the inset and the outset of the hill-top saddle cycle. We use Melnikov perturbation analysis, interpreted as an energy balance condition. This gets us the Melnikov result in the simplest way, using just the balance condition

energy input due to forcing = energy lost by dissipation.

These energies are calculated by integrating the work done by the forcing and damping as the system is imagined to move around the saddle-loop of the *Hamiltonian system*.

A.1   The Hamiltonian system

Setting $\beta = F = 0$ in equation (A.1), gives the Hamiltonian system

$$\ddot{x} + x - x^2 = 0,$$

which we regard as a mechanical oscillator with the energies

$$\text{kinetic energy} = T = \frac{1}{2}\dot{x}^2 = \frac{1}{2}y^2,$$
$$\text{potential energy} = V = \int(x - x^2)\,dx.$$

Hence, with an arbitrary constant C, we have

$$V = \frac{1}{2}x^2 - \frac{1}{3}x^3 + C.$$

The saddle-loop of the Hamiltonian system is shown in Fig. 31. It is symmetric about the x-axis. At a later stage of the analysis, it will be convenient to work with $s = x - 1$.

A.2   The energy loss due to dissipation

The incremental energy loss due to work done against the damping force is $\beta\dot{x}\delta x$ which we see is just β times the area of the hatched rectangle in Fig. 31. Thus, we have the useful result

total energy loss in the loop = $\beta \times$ area A.

Notice that we can evaluate this if we know $y(x)$. We do not need information about the variation of x or y with respect to time t.

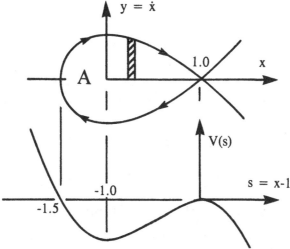

Figure 31: Hamiltonian saddle loop of the unforced and undamped escape equation. The top picture shows the phase trajectory while the lower picture shows the potential well.

A.3 The energy gain due to the forcing

The incremental work done by the forcing is $F \sin \omega(t + t_0) \, \delta x$ where the arbitrary t_0 allows us to vary the phase of the forcing in relation to the motion around the saddle-loop. So we have

$$\text{total energy gain in the loop} = \oint F \sin \omega(t + t_0) \, \dot{x} \, dt.$$

Notice that for this integral around the loop we need to know the time variation of $\dot{x}(t)$.

A.4 The phase-space loop and its area

The quickest way to obtain the equation of the Hamiltonian saddle-loop is to write

$$\text{total energy} = T + V = \text{constant} = \text{energy of hill-top equilibrium.}$$

In terms of $s \equiv x - 1$, with $\dot{x} = \dot{s} = y$, we have

$$T = \frac{1}{2}y^2,$$
$$V = -\frac{1}{6}s^2(3 + 2s).$$

Here, the arbitrary constant of V has been chosen so that the energy of the saddle equilibrium is zero, that is, $V = 0$ when $s = 0$. Hence, setting

$T + V = 0$, we have the equation of the saddle-loop

$$y = \pm s\sqrt{1 + 2s/3}.\qquad(A.2)$$

This is the $y(s)$ equation of the loop in phase space. Notice that the loop crosses the x-axis at $s = -\frac{3}{2}$ where V is also zero. Using the symmetry, the area of the loop is

$$A = 2\int_0^{-3/2} s\sqrt{1 + 2s/3}\,ds.$$

A standard integral quickly gives $A = \frac{6}{5}$ and so

$$\text{energy loss around the loop} = \frac{6\beta}{5}.\qquad(A.3)$$

A.5   Time variation along the Hamiltonian loop

For the energy gain, we need to determine the time variation around the loop by integrating (A.2). We rewrite the equation as

$$\int \frac{ds}{s\sqrt{1 + 2s/3}} = \pm\int dt.$$

Notice that the $\pm$ gives no trouble here, since it just stands in front of t, and we confirm that (A.4), (A.5) and (A.6) have the correct time flow. Using a second standard integral and taking the time datum such that the constant of integration is zero, gives us

$$s = \frac{3}{2}[\tanh^2(t/2) - 1],\qquad(A.4)$$

and, using the double angle formula, this simplifies to

$$s = \frac{-3}{1 + \cosh t},\qquad(A.5)$$

$$\dot{s} = \frac{3\sinh t}{(1 + \cosh t)^2}.\qquad(A.6)$$

We see that as t increases from $-\infty$ to zero, s decreases from zero to $-\frac{3}{2}$, and as t increases from zero to $+\infty$, s increases back to zero. So the time datum is centred symmetrically at $\dot{s} = 0$.

A.6   Evaluation of the energy gain

We can now write the energy gain E_G from the forcing as

$$E_G = \int_{-\infty}^{\infty} \frac{3F\sin\omega(t + t_0)\sinh t}{(1 + \cosh t)^2}\,dt.$$

This integral can be evaluated by the method of residues as

$$E_G = \frac{6F\pi\omega^2\cos\omega t_0}{\sinh\pi\omega}.\qquad(A.7)$$

A.7   The final energy balance

Equating the energy gain E_G of equation (A.7) to the energy loss E_L of equation (A.3) gives us the final energy balance condition

$$\frac{6F\pi\omega^2\cos\omega t_0}{\sinh\pi\omega} = \frac{6\beta}{5}.$$

Now we are trying to estimate the first tangency. In order to focus on the first occasion, under increasing F at constant β, at which a homoclinic orbit is possible, we must adjust the arbitrary relative phasing represented by t_0 to make F as small as possible. We can see that this is achieved by setting $t_0 = 0$ so that $\cos\omega t_0 = 1$, giving

$$\frac{F}{\beta} = \frac{\sinh\pi\omega}{5\pi\omega^2}.$$

We see that F is proportional to β, the constant of proportionality depending on the magnitude of ω. This equation is in good agreement with a numerically determined bifurcation locus (Foale and Thompson [7]).

The phasing of the homoclinic orbit is such that the forcing is just starting its sine-wave at the maximum excursion point where $\dot{s} = \dot{x} = 0$. This agrees with our homoclinic time-history (Fig. 19), on which the hilltop cycle will be almost exactly out of phase with the forcing. The Poincaré sampling on this figure is at 180°.

References

[1] Abraham, R.H. and Shaw, C.D., *Dynamics: The Geometry of Behaviour*, Addison-Wesley, Redwood City, 1992.

[2] Arrowsmith, D.K. and Place, C.M., *An Introduction to Dynamical Systems*, Cambridge University Press, Cambridge, 1990.

[3] Bishop, S.R. and Clifford, M.J., The use of manifold tangencies to predict orbits, bifurcations, and estimate escape in driven systems, *Proc. Euromech Coll.*, to appear, 1996.

[4] Clifford, M.J. and Bishop, S.R., Estimation of symmetry breaking and escape by observation of manifold tangencies, *Int. J. Bif. Chaos* 5, 883-890, 1995.

[5] Cusumano, J.P. and Kimble, B.W., Experimental observation of basins of attraction and homoclinic bifurcation in a magneto-mechanical oscillator. In *Nonlinearity and Chaos in Engineering Dynamics*, eds. J.M.T. Thompson and S.R. Bishop, Wiley, Chichester, 71-85, 1994.

[6] Falconer, K., *Fractal Geometry*, Wiley, Chichester, 1990.

[7] Foale, S. and Thompson, J.M.T., Geometrical concepts and computational techniques of nonlinear dynamics, *Comp. Meth. Appl. Mech. Engng* **89**, 381-394, 1991.

[8] Ghrist, R. and Holmes, P., Knots and orbit genealogies in three dimensional flows. In *Bifurcations and Periodic Orbits of Vector Fields*, NATO ASI Series, Kluwer, 1993.

[9] Glendinning, P., *Stability, Instability and Chaos*, Cambridge University Press, Cambridge, 1994.

[10] Gottwald, J.A., Virgin, L.N. and Dowell, E.H., Routes to escape from an energy well, *J. Sound Vib.* **187**, 133-144, 1995.

[11] Grebogi, C., Ott, E. and Yorke, J.A., Basin boundary metamorphoses: changes in accessible boundary orbits, *Physica D* **24**, 243-262, 1987.

[12] Guckenheimer, J. and Holmes, P., *Nonlinear Oscillations, Dynamical Systems and Bifurcations of Vector Fields*, Springer, New York, 1983.

[13] Kim, J.H. and Stringer, J. (eds.), *Applied Chaos*, Wiley, New York, 1992.

[14] Lansbury, A.N. and Thompson, J.M.T., Incursive fractals: a robust mechanism of basin erosion preceding the optimal escape from a potential well, *Phys. Letts A* **150**, 355-361, 1990.

[15] Lansbury, A.N., Thompson J.M.T. and Stewart, H.B., Basin erosion in the twin-well Duffing oscillator: two distinct bifurcation scenarios, *Int. J. Bif. Chaos* **2**, 505-532, 1992.

[16] MacMaster, A.G. and Thompson, J.M.T., Wave tank testing and the capsizability of hulls, *Proc. Roy. Soc. Lond. A* **446**, 217-232, 1994.

[17] McDonald, S.W., Grebogi, C., Ott, E. and Yorke, J.A., Fractal basin boundaries, *Physica D* **17**, 125-153, 1985.

[18] McRobie, F.A., Birkhoff signature change: a criterion for the instability of chaotic resonance, *Phil. Trans. Roy. Soc. Lond. A* **338**, 557-568, 1992.

[19] McRobie, F.A., Bifurcational precedences in the braids of periodic orbits of spiral 3-shoes in driven oscillators, *Proc. Roy. Soc. Lond. A* **438**, 545-569, 1992.

[20] McRobie, F.A. and Thompson, J.M.T., Lobe dynamics and the escape from a potential well, *Proc. Roy. Soc. Lond. A* **435**, 659-672, 1991.

[21] McRobie, F.A. and Thompson, J.M.T., Invariant sets of planar diffeomorphisms in nonlinear vibrations, *Proc. Roy. Soc. Lond. A* **436**, 427-448, 1992.

[22] McRobie, F.A. and Thompson, J.M.T., Braids and knots in driven oscillators, *Int. J. Bif. Chaos* **3**, 1343-1361, 1993.

[23] McRobie, F.A. and Thompson, J.M.T., Knot-types and bifurcation sequences of homoclinic and transient orbits of a single-degree-of-freedom driven oscillator, *Dyn. Stab. Sys.* **9**, 223-251, 1994.

[24] Mitsui, T., Ueda, Y. and Thompson, J.M.T., Straddle-orbit location of a chaotic saddle in a high-dimensional realisation of $\mathbf{R}^\infty$, *Proc. Roy. Soc. Lond. A* **445**, 669-677, 1994.

[25] Moon, F.C., *Chaotic and Fractal Dynamics*, Wiley, New York, 1992.

[26] Mullin, T. (ed.), *The Nature of Chaos*, Oxford University Press, Oxford, 1993.

[27] Parker, T.S. and Chua, L.O., *Practical Numerical Algorithms for Chaotic Systems*, Springer, New York, 1989.

[28] Rainey, R.C.T. and Thompson, J.M.T., The transient capsize diagram: a new method of quantifying stability in waves, *J. Ship Rsrch* **35**, 58-62, 1991.

[29] Schiehlen, W. (ed.), *Nonlinear Dynamics in Engineering Systems*, Proc. IUTAM Symp., Stuttgart, Springer, Berlin, 1990.

[30] Soliman, M.S., Predicting regimes of indeterminate jumps to resonance by assessing fractal boundaries in control space, *Int. J. Bif. Chaos* **4**, 1645-1653, 1994.

[31] Soliman, M.S. and Thompson, J.M.T., Integrity measures quantifying the erosion of smooth and fractal basins of attraction, *J. Sound Vib.* **135**, 453-475, 1989.

[32] Soliman, M.S. and Thompson, J.M.T., Basin organisation prior to a tangled saddle-node bifurcation, *Int. J. Bif. Chaos* **1**, 107-118, 1991.

[33] Soliman, M.S. and Thompson, J.M.T., Global dynamics underlying sharp basin erosion in nonlinear driven oscillators, *Phys. Rev. A* **45**, 3425-3431, 1992.

[34] Soliman, M.S. and Thompson, J.M.T., Indeterminate subcritical bifurcations in parametric resonance, *Proc. Roy. Soc. Lond. A* **438**, 511-518, 1992.

[35] Soliman, M.S. and Thompson, J.M.T., Indeterminate transcritical bifurcations in parametrically excited systems, *Proc. Roy. Soc. Lond. A* **439**, 601-610, 1992.

[36] Soliman, M.S. and Thompson, J.M.T., Indeterminate bifurcational phenomena in hardening systems, *Proc. Roy. Soc. Lond. A*, in press, 1996.

[37] Stewart, H.B., A chaotic saddle catastrophe in forced oscillators. In *Dynamical Systems Approaches to Nonlinear Problems in Systems and Circuits*, eds. F. Salam and M. Levi, SIAM, Philadelphia, 1987.

[38] Stewart, H.B., Thompson, J.M.T., Ueda Y. and Lansbury, A.N., Optimal escape from potential wells: patterns of regular and chaotic bifurcation, *Physica D* **85**, 259-295, 1995.

[39] Stewart, H.B. and Ueda, Y., Catastrophes with indeterminate outcome, *Proc. Roy. Soc. Lond. A* **432**, 113-123, 1991.

[40] Thompson, J.M.T., Chaotic phenomena triggering the escape from a potential well, *Proc. Roy. Soc. Lond. A* **421**, 195-225, 1989.

[41] Thompson, J.M.T., Global unpredictability in nonlinear dynamics: capture, dispersal and the indeterminate bifurcations, *Physica D* **58**, 260-272, 1992.

[42] Thompson, J.M.T., Basic concepts of nonlinear dynamics. In *Nonlinearity and Chaos in Engineering Dynamics*, eds. J.M.T. Thompson and S.R. Bishop, Wiley, Chichester, 1-21, 1994.

[43] Thompson, J.M.T. and Chua, L.O. (eds.), *Chaotic Behaviour in Electronic Circuits*, Theme Issue, *Phil. Trans. Roy. Soc. Lond. A* **353**, 1-136, 1995.

[44] Thompson, J.M.T. and Gray, P. (eds.), *Chaos and Dynamical Complexity in the Physical Sciences*, Theme Issue, *Phil. Trans. Roy. Soc. Lond. A* **332**, 49-186, 1990.

[45] Thompson, J.M.T. and McRobie, F.A., Indeterminate bifurcations and the global dynamics of driven oscillators. In *First European Nonlinear Oscillations Conf., Hamburg*, eds. E. Kreuzer and G. Schmidt, Akademie Verlag, Berlin, 107-128, 1993.

[46] Thompson, J.M.T., Rainey, R.C.T. and Soliman, M.S., Ship stability criteria based on chaotic transients from incursive fractals, *Phil. Trans. Roy. Soc. Lond. A* **332**, 149-167, 1990.

[47] Thompson, J.M.T., Rainey, R.C.T. and Soliman, M.S., Mechanics of ship capsize under direct and parametric wave excitation, *Phil. Trans. Roy. Soc. Lond. A* **338**, 471-490, 1992.

[48] Thompson, J.M.T. and Schiehlen, W. (eds.), *Nonlinear Dynamics of Engineering Systems*, Theme Issue, *Phil. Trans. Roy. Soc. Lond. A* **338**, 451-568, 1992.

[49] Thompson, J.M.T. and Soliman, M.S., Fractal control boundaries of driven oscillators and their relevance to safe engineering design, *Proc. Roy. Soc. Lond. A* **428**, 1-13, 1990.

[50] Thompson, J.M.T. and Soliman, M.S., Indeterminate jumps to resonance from a tangled saddle-node bifurcation, *Proc. Roy. Soc. Lond. A* **432**, 101-111, 1991.

[51] Thompson, J.M.T. and de Souza, J.R., Suppression of escape by resonant modal interactions: in shell vibration and heave-roll capsize, *Proc. Roy. Soc. Lond. A*, submitted, 1996.

[52] Thompson, J.M.T. and Stewart, H.B., *Nonlinear Dynamics and Chaos*, Wiley, Chichester, 1986.

[53] Thompson, J.M.T. and Stewart, H.B., A tutorial glossary of geometrical dynamics, *Int. J. Bif. Chaos* **3**, 223-239, 1993.

[54] Thompson, J.M.T., Stewart, H.B. and Ueda, Y., Safe, explosive and dangerous bifurcations in dissipative dynamical systems, *Phys. Rev. E* **49**, 1019-1027, 1994.

[55] Thompson, J.M.T. and Ueda, Y., Basin boundary metamorphoses in the canonical escape equation, *Dyn. Stab. Syst.* **4**, 285-294, 1989.

[56] Ueda, Y., Yoshida, S., Stewart, H.B. and Thompson, J.M.T., Basin explosions and escape phenomena in the twin-well Duffing oscillator: compound global bifurcations organising behaviour, *Phil. Trans. Roy. Soc. Lond. A* **332**, 169-186, 1990.

An Introduction to the Mechanisms and Methods for the Detection of Chaos

S. Wiggins

Departments of Applied Mechanics and
Control and Dynamical Systems 104-44,
Caltech,
Pasadena,
U.S.A.

1 Chaos: What is It and How Does It Occur?

In this article we give an introduction to the essential features of the type of chaotic dynamics associated with homoclinic and heteroclinic orbits. In particular, we will discuss the following issues.

- What is meant by the term "chaos" in this context?

- How do homoclinic and heteroclinic orbits give rise to this type of chaos?

- What are the implications of this type of chaos in specific systems where it arises?

It takes awhile to get through this discussion. We will begin by discussing the shift map. This is a dynamical system that in some sense can be viewed as a paradigm for chaotic dynamics. However, to an engineer or scientist, this type of dynamical system hardly seems practical in the sense that it does not appear possible that it would occur in the types of dynamical systems arising in applications. We address this by considering a simple two-dimensional map that contains the essential ingredients that give rise to the shift map dynamics. This is the Smale horseshoe map. But again, even though it is a map defined on a typical type of phase space that arises in applications ($\mathbf{R}^2$ as opposed to the space of bi-infinite sequences of 0's and 1's for the shift map), the features of the horseshoe map that are responsible for creating the shift map dynamics appear to be very delicate and contrived solely for that purpose (which they were). Nevertheless, they are much more robust than one might believe at first glance, and this is the content of the so-called "Conley-Moser" conditions. But even after all this, it is not at all clear how one would take a concrete dynamical system (e.g. a set of differential equations arising through a modelling procedure in some applications) and show that the situation is such that the Conley-Moser conditions could be verified, which would imply Smale horseshoe like dynamics, which would imply shift map type chaos. Finally, in answer to this, we show how certain types of solutions, namely homoclinic orbits and heteroclinic cycles, *may* give rise to a situation where the Conley-Moser conditions can be verified, with all the attendant consequences. This latter step is important because we can then develop analytical methods for detecting when these types of solutions arise in specific dynamical systems, which is the subject of the latter part of this article.

We summarise this chain of reasoning below.

```
┌─────────────────────────────────────────────────────────────┐
│  Does this type of dynamics arise in "real" dynamical systems? │
└─────────────────────────────────────────────────────────────┘
                              ↓
                     The Smale Horseshoe
                              ↓
┌─────────────────────────────────────────────────────────────┐
│     Yes, but how general is the horseshoe construction?       │
└─────────────────────────────────────────────────────────────┘
                              ↓
                  The Conley-Moser conditions
                              ↓
┌─────────────────────────────────────────────────────────────┐
│  But could you ever verify these conditions for an explicit example? │
└─────────────────────────────────────────────────────────────┘
                              ↓
            Homoclinic Orbits and Heteroclinic Cycles:
                    A Jungle of Possibilities
                              ↓
┌─────────────────────────────────────────────────────────────┐
│ Global, Geometric Perturbation Methods for Detecting Homoclinic Orbits │
└─────────────────────────────────────────────────────────────┘
```

1.1   The shift map: the paradigm for chaos

The phase space for the shift map Σ is the space of "bi-infinite" sequences of 0's and 1's:

$$s \in \Sigma, \quad s = \{ \ldots s_{-n} \ldots s_{-1}.s_0 s_1 \ldots s_n \ldots \}, \qquad s_i = 0 \text{ or } 1.$$

We define the following metric on Σ:

$$d(s, \bar{s}) = \sum_{i=-\infty}^{\infty} \frac{\delta_i}{2^{|i|}}, \qquad \text{where} \quad \delta_i = \begin{cases} 0 & \text{if } s_i = \bar{s}_i, \\ 1 & \text{if } s_i \neq \bar{s}_i. \end{cases}$$

Two sequences are "near each other" according to this metric if they are identical on a long central block (the longer the closer, see Wiggins [17], Lemma 4.2.2). It can be proved that Σ is compact, perfect (i.e. every point is a limit point), and totally disconnected, i.e. Σ is a Cantor set. Moreover, it has the cardinality of the continuum (see Wiggins [17], Proposition 4.2.4).

The *shift map* σ acting on Σ is defined as follows:

$$\sigma : \Sigma \longrightarrow \Sigma,$$

$$\sigma(s) = \{\ldots s_{-n} \ldots s_{-1} s_0 . s_1 s_2 \ldots s_n \ldots\},$$

$$\text{or} \qquad [\sigma(s)]_i = s_{i+1}.$$

It is easy to verify that σ is a homeomorphism and that it has the following properties:

1. a countable ∞ of periodic orbits of all periods,

2. an uncountable ∞ of nonperiodic orbits, and

3. a "dense orbit", i.e. an orbit that is dense in Σ.

(These properties are proved in Wiggins [17], Proposition 4.2.7.)

The dynamics of $\sigma : \Sigma \longrightarrow \Sigma$ model the phenomenon that we refer to as *deterministic chaos* in that it has the following ingredients:

- Σ is compact;

- σ is topologically transitive, i.e. given any two open sets in Σ some iterate of one will intersect the other (this essentially follows from the existence of the dense orbit);

- σ exhibits sensitive dependence on initial conditions, i.e. the distance between nearby initial conditions grows under some fixed number of iterates.

(Precise definitions and proofs can be found in Wiggins [17].) Again, proving such characteristics for maps in general is difficult. It is the special structure of this dynamical system that makes the proofs trivial.

Practically speaking, the shift map models exactly what we would like to embody in the idea of deterministic chaos, that is:

> *despite the fact that the system is deterministic, it has the property that imprecise knowledge of the initial condition may lead to unpredictability after some finite time.*

For the shift map one sees this as follows. Specifying the initial condition for this dynamical system means specifying a bi-infinite sequence. However, in practice (e.g. in engineering or scientific applications) one only knows the initial sequence up to some error. Using our metric, this means that we only know the exact value of the sequence for some long central block, say N entries to the right of the decimal point and N to the left. Hence after N forward iterations by σ the orbit has "forgotten" its initial condition, since the sequence is unknown from that point on. Consequently, its future can no longer be predicted.

The shift map acting on the space of bi-infinite symbol sequences hardly seems like the typical dynamical system arising in applications. Next we want to move one step closer to this by considering the Smale horseshoe map.

1.2   The Smale horseshoe

The Smale horseshoe map is a two-dimensional map, denoted f, of the square with sides of length one, denoted D, into the plane. It is defined by a combination of analytical and geometrical properties. Roughly speaking, f contracts the square in the horizontal direction, expands it in the vertical direction, and folds it on to itself as shown in Fig. 1. In particular, the two "horizontal" rectangles H_0 and H_1 are mapped to the two vertical rectangles V_0 and V_1 respectively, as shown in Fig. 1. Moreover, horizontal boundaries of H_i map to horizontal boundaries of V_i, $i = 0, 1$ and vertical boundaries of H_i map to vertical boundaries of V_i, $i = 0, 1$. (This is not a mere technical point, the following would not be true otherwise.)

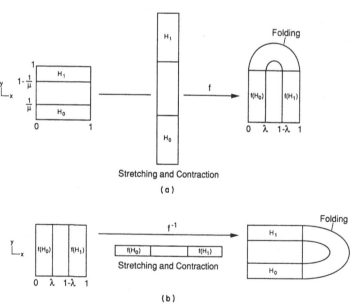

Figure 1: Action of the Smale horseshoe map on horizontal and vertical rectangles.

On these horizontal rectangles the map is analytically given by

$$f|_{H_0} : \begin{pmatrix} x \\ y \end{pmatrix} \longmapsto \begin{pmatrix} \lambda & 0 \\ 0 & \mu \end{pmatrix} \begin{pmatrix} x \\ y \end{pmatrix},$$

$$f|_{H_1} : \begin{pmatrix} x \\ y \end{pmatrix} \longmapsto \begin{pmatrix} -\lambda & 0 \\ 0 & -\mu \end{pmatrix} \begin{pmatrix} x \\ y \end{pmatrix} + \begin{pmatrix} 1 \\ \mu \end{pmatrix},$$

where $0 < \lambda < \frac{1}{2}$ and $\mu > 2$. The folding is a result of the nonlinearity and occurs in the region between H_0 and H_1. The precise analytical form of the map in this region is not important for our purposes. Rather, the important feature is that the map behaves qualitatively as described.

Next we consider the invariant set of this map

$$\Lambda = \bigcap_{n=-\infty}^{\infty} f^n(D).$$

Λ is the set of points in the square that remain in the square under all forward and backward iterations by f.

Using the definition of f given above (and in particular the fact that "horizontal" boundaries map to "horizontal" boundaries and "vertical" boundaries map to "vertical" boundaries, with $\lambda\mu < 1$), we can recursively construct the set in several steps. We define

$$\Lambda_-^n \equiv D \cap f(D) \cap f^2(D) \cap \ldots \cap f^n(D),$$

and show that it consists of 2^n vertical rectangles each of width λ^n. This result can be understood as follows. The action of f is to contract the square by a factor λ in the horizontal direction, expand it by a factor μ in the vertical direction, fold the result, and lay it back down on the square, where it is layed down respecting the original horizontal and vertical boundaries. The result is that $D \cap f(D)$ consists of two rectangles, each having vertical sides of length one and horizontal sides of length λ. Now apply f to the square again, but track the two rectangles, i.e. $D \cap f(D)$, in the process. They get squeezed together in the horizontal direction (with their widths contracted by another factor of λ), expanded in the vertical direction by a factor μ, folded and layed back down on the square respecting horizontal and vertical boundaries. The result is that $f(D \cap f(D)) \cap D$ consists of four rectangles, each of vertical length one and horizontal width λ^2. This process can can be carried out indefinitely. After each iterate the number of rectangles is doubled, with their horizontal widths decreased by a factor λ (and remember $\lambda < 1$).

Passing to the limit as $n \to \infty$ gives

$$\Lambda_- \equiv D \cap f(D) \cap f^2(D) \cap \ldots \cap f^n(D) \cap \ldots,$$

which is a Cantor set of vertical lines. These vertical lines are the set of points in D that remain in D under all *backward* iterations of f.

Similarly, for any n the set

$$\Lambda_+^n \equiv D \cap f^{-1}(D) \cap f^{-2}(D) \cap \ldots \cap f^{-n}(D),$$

consists of 2^n horizontal rectangles each of width $\frac{1}{\mu^n}$. Passing to the limit as $n \to \infty$ gives

$$\Lambda_+ \equiv D \cap f^{-1}(D) \cap f^{-2}(D) \cap \ldots \cap f^{-n}(D) \cap \ldots,$$

which is a Cantor set of horizontal lines, which have the interpretation as the set of points in D that remain in D under all *forward* iterations of f. Hence the set

$$\Lambda \equiv \Lambda_+ \cap \Lambda_-$$

is a Cantor set of *points*. These are the points in D that remain in D under all iterations by f.

1.2.1   Dynamics on the invariant set

The dynamics on the invariant set of the Smale horseshoe map is "modelled" by the shift map. By this we mean that there is a change of coordinates that transforms the Smale horseshoe map, *restricted to* Λ, to $\sigma : \Sigma \to \Sigma$. This "change of coordinates" is defined as follows:

$$\phi : \Lambda \longrightarrow \Sigma,$$
$$p \longmapsto \phi(p) = \{ \ldots s_{-n} \ldots s_{-1}.s_0 s_1 \ldots s_n \ldots \},$$

where

$$s_i = \begin{cases} 0 & \text{if } f^i(p) \in H_0, \\ 1 & \text{if } f^i(p) \in H_1. \end{cases}$$

It can be shown that ϕ is a homeomorphism (see Wiggins [17], Theorem 4.1.3). By construction we have

$$\sigma \circ \phi(p) = \phi \circ f(p).$$

Since ϕ is a homeomorphism

$$\phi^{-1} \circ \sigma \circ \phi = f,$$

and therefore

$$\phi^{-1} \circ \sigma^n \circ \phi = f^n.$$

Thus, every orbit of the shift map is mapped uniquely to an orbit of $f|_\Lambda$. In particular, $f|_\Lambda$ has the same orbit structure as the shift map: a countable infinity of periodic orbits of all periods, an uncountable infinity of non-periodic orbits and an orbit that is dense in Λ. It is also *chaotic* in the same sense as we described for the shift map.

1.3   The Conley-Moser conditions, or "How you prove that 'horseshoe-like' dynamics exist in an explicit example"

One might still argue that Smale's horseshoe map is rather artificial from the point of view of the types of maps arising in applications. From a traditional applied mathematics point of view this may be true, but from the geometrical dynamical systems point of view the Smale horseshoe map contains the essential characteristics that give rise to the shift map dynamics in many applications. These were explicitly brought out by Conley and Moser. They found conditions for a general two dimensional map to satisfy in order for it to have an invariant set on which it has the shift dynamics. These conditions are a combination of geometrical and analytical conditions. The geometrical part consists of generalising the notion of horizontal and vertical rectangles by allowing the boundaries to be Lipschitz curves, rather than straight lines. With this generalisation in hand one then requires horizontal "rectangles" to map to "vertical" rectangles with horizontal boundaries mapping to horizontal boundaries and vertical boundaries mapping to vertical boundaries. The analytical part comes from requiring uniform (but not constant) contraction in the horizontal directions and expansion in the vertical direction (see Fig. 2). Details on the Conley-Moser conditions for two-dimensional maps are in Wiggins [17], the n-dimensional case is discussed in Wiggins [16].

From the point of view of applications, the significance of these conditions is that they are satisfied in the neighbourhood of certain orbits of maps or differential equations. We next describe this situation.

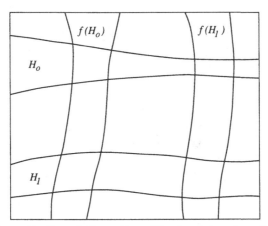

Figure 2: Generalisation of the notion of horizontal and vertical rectangles for the Conley-Moser conditions.

1.4   What is a homoclinic orbit and how does it give rise to this behaviour?

A homoclinic orbit is an orbit that is both forward and backward asymptotic to the same invariant set. Hence there are two parts to the definition: an invariant set and asymptotic behaviour with respect to this invariant set. Each of these parts play a distinct role in the possibility of a homoclinic orbit giving rise to chaotic dynamics.

Various types of invariant sets are possible. For example,

- an equilibrium point,

- a periodic orbit,

- a quasiperiodic orbit (or invariant torus),

- a normally hyperbolic invariant manifold,

- a general invariant set.

The characteristic of the orbit being forward and backward asymptotic to the invariant set is described (and more quantitatively dealt with) by saying that the orbit is in both the *stable* and *unstable manifolds* of the invariant set, or that the stable and unstable manifolds of the invariant set *intersect*. Two issues arise in this context:

- transversality of the intersection,

- global geometry of the intersection.

Next we will consider some specific invariant sets and show how Smale horseshoes can be constructed near the intersections of the stable and unstable manifolds of these invariant sets.

1.4.1   Orbits homoclinic to a periodic orbit

This is one of the most general situations. Consider a map or an ordinary differential equation having a saddle-type (hyperbolic) periodic orbit whose stable and unstable manifolds intersect transversely. (The ordinary differential equation setting can be converted to the map setting by passing to the so-called Poincaré map defined in a neighbourhood of the periodic orbit.) Then near a point on the resulting homoclinic orbit one can find an invariant set on which the system is topologically conjugate to the shift map. This is the content of the Smale-Birkhoff Theorem or Moser's Theorem (see Wiggins [17], Section 4.4 for a discussion). The saddle-type nature of the periodic orbit gives rise to the stretching and contraction of phase space. The transversality of the intersection of the stable and unstable manifolds provides coordinate directions associated with which one can find horizontal rectangles that map to vertical rectangles with horizontal (resp. vertical) boundaries mapping to horizontal (resp. vertical) boundaries. Thus, a region near a point of transverse intersection (a *transverse homoclinic point*) of the stable and unstable manifolds can be found on which the Conley-Moser conditions can be verified. We illustrate this in Fig. 3 where we show how a region maps around in the homoclinic tangle back over itself. In Fig. 4 we pick out the horizontal and vertical strips.

This result is true in arbitrary, but finite, dimensions and some infinite dimensional results are also known.

1.4.2 What does the "shift map"-type chaos mean?

What does the existence of an invariant Cantor set on which the dynamics is topologically conjugate to the shift map imply for a particular example? The point we want to make here is that answering this question requires knowledge about the geometry of the homoclinic orbits and its interpretation in terms of the dynamics of the particular example.

For example, consider a "particle in a double-well potential" subject to periodic forcing, e.g. the periodically forced Duffing equation

$$\begin{aligned}
\dot{x} &= y, \\
\dot{y} &= x - x^3 + f \cos \omega t, \qquad (x, y) \in \mathbf{R}^2.
\end{aligned}$$

For $f = 0$ the phase portrait appears as in Fig. 5. There are three families of periodic orbits, one in each well and one encircling both wells. The three families are separated by a pair of homoclinic orbits. For $f \neq 0$, in the associated Poincaré map, the homoclinic orbits "break", and intersect each other transversely giving rise to the familiar homoclinic tangle (see Fig. 6). In this case there is a mechanism for trajectories starting in one well to enter the other, and vice-versa. Now the Smale-Birkhoff Theorem or Moser's Theorem immediately implies that near the transverse intersection points of the stable and unstable manifolds of the saddle there exists an invariant set on which the dynamics is topologically equivalent to the shift map. But what does that mean in terms of chaos for a particle moving in a double-well potential subject to periodic excitation? In this situation one can use the Conley-Moser conditions to construct symbolic dynamics so that a 0 corresponds to motion in the right well and a 1 corresponds to motion in the left well. Hence the chaos in this case would correspond to "random" jumping from well to well.

1.4.3 Heteroclinic cycles

A heteroclinic orbit is an orbit lying in the intersection of the stable and unstable manifold of two *different* invariant sets. In this case it is not possible

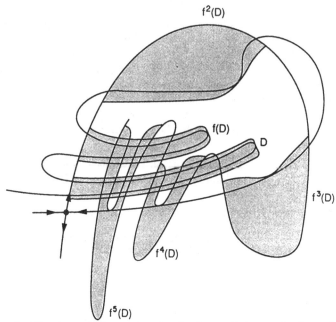

Figure 3: The image of a region in the homoclinic tangle formed by the transverse intersection of the stable and unstable manifolds of a hyperbolic saddle point.

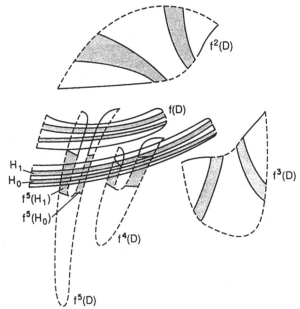

Figure 4: The image of two select horizontal rectangles within this region.

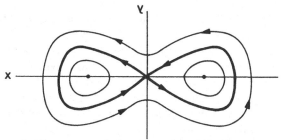

Figure 5: Phase portrait for the unforced Duffing equation.

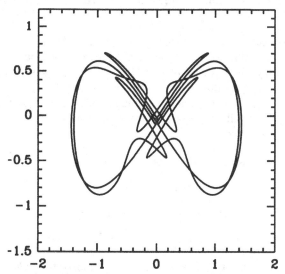

Figure 6: Homoclinic tangle in the forced Duffing equation.

(unless the dynamical system was on the cylinder, torus, or sphere) to construct something like a Smale horseshoe near a heteroclinic point. The reason
for this is that in forward and backward time points near a heteroclinic point
move to different regions of the phase space, i.e. we cannot find horizontal
strips that are contracted, expanded, and mapped over themselves so that the
Conley-Moser conditions can be applied. However, the case of a heteroclinic
cycle is different.

We discuss heteroclinic cycles for maps. Consider a map $f : \mathbf{R}^n \to \mathbf{R}^n$
having hyperbolic fixed points

$$p_0, p_1, p_2, \ldots, p_n \equiv p_0,$$

$$W^u(p_i) \cap W^s(p_{i+1}), \quad i = 0, \ldots, n-1.$$

Then if the intersections of the stable and unstable manifolds are transverse, it
can be proved that this implies the existence of transverse homoclinic points

(see Palis & de Melo [8], p85). Hence, the Smale-Birkhoff Theorem and Moser's Theorem apply.

Note that there are many other possibilities besides heteroclinic cycles formed by the intersection of the stable and unstable manifolds of hyperbolic fixed points of maps. One could consider heteroclinic cycles formed by the intersection of stable and unstable manifolds of different types of invariant sets such as equilibrium points, periodic orbits, and invariant tori. Very little work has been done in this area.

1.4.4   Orbits homoclinic to equilibria of ordinary differential equations

For transverse intersections of the stable and unstable manifolds of a hyperbolic periodic orbit there are the general theorems of Smale-Birkhoff and Moser that imply the existence of chaos in the sense of the shift map. These theorems are very general. However, when we consider the case of orbits homoclinic to equilibrium points of ordinary differential equations there is not such a general, all encompassing theorem. Rather, there is a jungle of possibilities depending on

- eigenvalues associated with the fixed point,

- dimension of the phase space,

- global geometry of the stable and unstable manifolds,

- special structure (e.g. symmetries, Hamiltonian, etc.).

We give two examples in three dimensions that illustrate the key issues.

Sil'nikov's example.   We first describe the famous example of Sil'nikov [11]. Consider the following situation:

- an autonomous vector field in three dimensions,

- it possesses an equilibrium point with eigenvalues associated with the linearisation about the equilibrium point given by $\rho \pm i\omega$ and λ where $\rho < 0$, $\lambda > 0$, $\omega \neq 0$ with

$$\lambda > -\rho > 0,$$

- the equilibrium point is connected to itself by a homoclinic orbit.

The standard approach to the analysis of the dynamics near this homoclinic orbit is to construct a Poincaré map in the neighbourhood of the homoclinic orbit (which, of course, is not defined *on* the homoclinic orbit). This map consists of the composition of two maps. One is a map given by trajectories

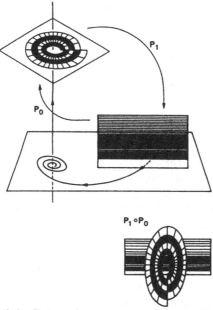

Figure 7: Action of the Poincaré map on a region near the homoclinic orbit in Sil'nikov's example.

as they pass near the equilibrium point (denoted P_0) and the other is given by trajectories near the homoclinic orbit (denoted P_1) as they pass near the homoclinic orbit outside a neighbourhood of the equilibrium point. In Fig. 7 we show the action of this Poincaré map on a region near the homoclinic orbit. In the passage near the equilibrium point it is stretched, contracted and folded around the homoclinic orbit. The passage along the homoclinic orbit serves to map this region back on top of itself. In this stretching, contracting and folding process one can find horizontal strips in the original region that are stretched, contracted and mapped back over themselves in a way that respects horizontal and vertical boundaries. Hence, the Conley-Moser conditions can be verified and we can conclude that there is an invariant set on which the dynamics is topologically conjugate to the shift dynamics.

"Lorenz equation like" example. Next we consider an example much like the Lorenz equations (Sparrow [13]). Consider the following situation:

- an autonomous vector field in three-dimensions,

- it possesses an equilibrium point with purely real eigenvalues given by

$$0 < -\lambda_2 < \lambda_3 < -\lambda_1, \qquad \lambda_i > 0, \quad i = 1, 2, 3,$$

- the equilibrium point is connected to itself by a *pair* of homoclinic orbits.

The analysis of the dynamics near the homoclinic orbit is carried out just as in the case for Sil'nikov's example. In Fig. 8 we show a cross-section to the flow. The line in the middle represents the intersection of the two-dimensional stable manifold of the origin with this cross-section. We choose two "horizontal rectangles" on each side of this line and consider their images under the Poincaré return map. As they pass near the origin they experience stretching and contraction (as shown by their transformations into the "half-bowtie" shapes in Fig. 9), *but no folding*, since the eigenvalues are purely real. As these "half-bowties" travel along their respective homoclinic orbits outside a neighbourhood of the origin they return to the cross-section as shown in Fig. 9. We see that any one "half-bowtie" is not mapped over *both* horizontal rectangles in its return to the cross-section. However, this situation will change if we have at our disposal a parameter to vary that will "break" the homoclinic connection. In particular, if it causes the homoclinic orbits to move in such a manner that each half-bowtie maps over both horizontal rectangles as shown in Fig. 10. In this case we can find horizontal rectangles that are stretched, contracted and mapped over themselves in a way that the Conley-Moser conditions can be verified. Hence, there is an invariant set on which the dynamics is topologically conjugate to the shift dynamics.

Now let us compare this situation with Sil'nikov's example.

1. In Sil'nikov's example the shift dynamics (or "Smale horseshoes", or chaos) occurs when there is a single homoclinic orbit whereas, in the "Lorenz-like" example, two homoclinic orbits are required.

2. In Sil'nikov's example the shift dynamics is present when the homoclinic orbit exists whereas, in the "Lorenz-like" example, the homoclinic orbits have to be "broken" such as through varying a parameter.

Each of these differences arises from the same fact. Namely, that a pair of the eigenvalues in Sil'nikov's example have nonzero complex parts and the eigenvalues in the "Lorenz equation like" example are purely real. Roughly speaking, the spiraling associated with the complex eigenvalues twists the image of the cross-section under the Poincaré map around the homoclinic orbit (this is where the folding arises). If the eigenvalues are purely real there is no spiraling and a pair of homoclinic orbits is required to effect the folding process. Moreover, in this case the homoclinic orbit must be "broken" so that the image of one horizontal rectangle maps over both horizontal rectangles on its return to the Poincaré section.

Higher dimensional versions of these results hold as well as results for Hamiltonian systems (Wiggins [16] summarises these results for two degree-of-freedom Hamiltonian systems). Some additional recent references on homo-clinic and heteroclinic orbits are Arnol'd, Afrajmovich, Il'yashenko & Sil'nikov [2], Palis & Takens [9], *Physica D*, Vol 62, 1993. This latter reference contains a nice compilation of applications where homoclinic phenomena arise.

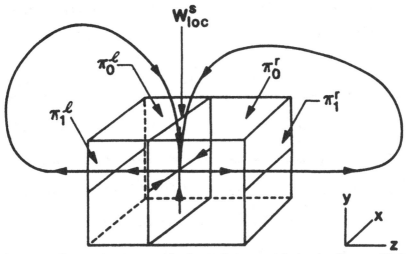

Figure 8: Cross-sections near the homoclinic example in the "Lorenz-like" example.

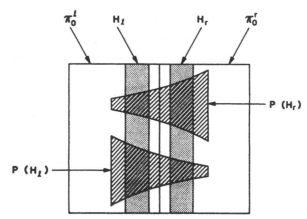

Figure 9: Images of horizontal rectangles in the cross-section under the Poincaré map near the homoclinic orbit.

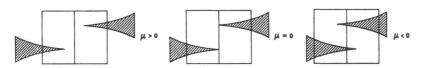

Figure 10: The "Smale horseshoe like" behaviour in the Poincaré map near the homoclinic orbit as the parameter is varied.

2   A Normal Form for Global Perturbation Methods

In Section 1 we discussed homoclinic orbits as a mechanism for chaotic dynamics. In this section we will begin the development of some analytical techniques for detecting these mechanisms in specific dynamical systems. The techniques will be developed in the context of a specific class of dynamical systems which can be viewed as a normal form. More details can be found in Wiggins [16], Chapter 4.

The class of dynamical systems that we will consider is of the form

$$
\begin{aligned}
\dot{x} &= JD_xH_0(x,I) + \epsilon g^x(x,I,\theta,\mu,\epsilon), \\
\dot{I} &= \epsilon g^I(x,I,\theta,\mu,\epsilon), \\
\dot{\theta} &= \Omega(x,I) + \epsilon g^\theta(x,I,\theta,\mu,\epsilon),
\end{aligned}
$$

where $x \in \mathbf{R}^2$, $J = \begin{pmatrix} 0 & 1 \\ -1 & 0 \end{pmatrix}$, $I \in \mathbf{R}^m$ or T^m. Also $\theta \in T^l$ and $\mu \in \mathbf{R}^p$ are parameters, and $0 < \epsilon \ll 1$ is the perturbation parameter. We remark that we can treat the case $x \in \mathbf{R}^{2n}$, $n > 1$, but that creates some complications which we do not wish to address here. We will assume that the functions are sufficiently differentiable for our needs. This normal form encompasses many types of systems that arise in applications. Below we give some examples.

2.1   Examples

Two Degree-of-Freedom "Near Integrable" Hamiltonian Systems.
Consider the system

$$
\begin{aligned}
\dot{x} &= JD_xH_0(x,I) + \epsilon\left(JD_xH_1(x,I,\theta) + \bar{g}^x(x,I,\theta)\right), \\
\dot{I} &= \epsilon\left(-D_\theta H_1(x,I,\theta) + \bar{g}^I(x,I,\theta)\right), \\
\dot{\theta} &= D_IH_0(x,I) + \epsilon\left(D_IH_1(x,I,\theta) + \bar{g}^\theta(x,I,\theta)\right),
\end{aligned}
$$

where $x \in \mathbf{R}^2$, $I \in \mathbf{R}$ and $\theta \in T^1$. For $\epsilon = 0$ this system is a completely integrable two-degree-of-freedom Hamiltonian system. This is easily seen since I is constant, and the x component is therefore a one-parameter (with I as the parameter) family of one-degree-of-freedom Hamiltonian systems; hence the x component can be solved for separately. Once this is done the θ component is obtained by quadrature. We then consider perturbations to this system which may be of two types, Hamiltonian and/or non-Hamiltonian.

Slowly Varying Oscillators. Consider the system

$$
\dot{x} = JD_xH_0(x,I) + \epsilon g^x(x,I,\theta),
$$

$$\dot{I} = \epsilon g^I(x, I, \theta),$$
$$\dot{\theta} = \omega,$$

where $x \in \mathbf{R}^2$, $I \in \mathbf{R}$, $\theta \in T^1$ and ω is a constant. The x component can be viewed as a periodically forced one-degree-of-freedom oscillator depending on the slowly varying parameter I. The periodic forcing comes from the fact that $\theta(t) = \omega t + \theta_0$. We have included the phase as a dependent variable in order to make the system autonomous. This has some conceptual utility, e.g. it aids in the construction of Poincaré maps. Moreover, I obeys its own first order dynamical equation.

Quasiperiodically Forced Oscillators. Consider the system

$$\dot{x} = JD_x H_0(x, I) + \epsilon g^x(x, \theta),$$
$$\dot{\theta} = \omega,$$

where $x \in \mathbf{R}^2$, $\theta \in T^l$ and ω is a constant. Here $\theta(t) = \omega t + \theta_0$. However, as opposed to the previous case θ is a vector, not a scalar. Hence this system has the form of a quasiperiodically forced oscillator.

Adiabatically Forced Oscillators. Consider the system

$$\dot{x} = JD_x H_0(x, I),$$
$$\dot{I} = \epsilon \omega,$$

where $x \in \mathbf{R}^2$, $I \in T^1$ and ω is a constant. This one-degree-of-freedom system depends periodically on a slowly varying parameter. However, its character is very different from that of the examples described above. In the above examples the perturbation only modifies the vector field by an $\mathcal{O}(\epsilon)$ amount. In this example in the course of the time evolution through one period $\frac{2\pi}{\epsilon\omega}$, the vector field may undergo an $\mathcal{O}(1)$ change.

2.2   Geometry of the unperturbed normal form

We refer to the perturbation methods that we develop as *geometrical perturbation methods*. The reason for this is that the geometrical structures of the unperturbed system in phase space form the backbone or "skeleton" on which the analytical methods for studying the perturbed system are developed. For this reason we first describe the geometrical structure of the unperturbed system, i.e. the system obtained by setting $\epsilon = 0$:

$$\dot{x} = JD_x H_0(x, I),$$
$$\dot{I} = 0, \tag{2.1}$$
$$\dot{\theta} = \Omega(x, I),$$

where $x \in \mathbf{R}^2$, $I \in \mathbf{R}^m$ or T^m and $\theta \in T^l$. We make an analytical assumption on the x component from which all the geometrical consequences follow.

Assumption on the x-Component: For all $I \in B \subset \mathbf{R}^m$, where B is some open set (or for all $I \in T^m$), $\dot{x} = JD_xH_0(x, I)$ has a hyperbolic fixed point $x_0(I)$ that is connected to itself by a homoclinic orbit $x^h(t, I)$.

It follows from this assumption that in the full phase space the set

$$\mathcal{M}_0 = \{(x, I, \theta) | x = x_0(I)\} \tag{2.2}$$

is an $m + l$ dimensional *normally hyperbolic* invariant manifold that is as differentiable as the vector field. How do we see this? A manifold is said to be invariant if trajectories starting on the manifold stay on it forever, or if they leave the surface they only leave by crossing its boundary. Equivalently, a manifold is invariant if the vector field is tangent to the manifold. This latter condition is analytical in nature and, since we have explicit formulae for the manifold and the vector field, can be verified by a calculation, which we now do.

From (2.2), $\mathcal{M}_0$ is given by the graph of the function

$$x = x_0(I).$$

The condition for tangency of the vector field to the manifold can be verified by substituting a trajectory of (2.1) into the expression for $\mathcal{M}_0$ and differentiating the result with respect to t (time). Tangency follows if the resulting expression is zero. In our case we have

$$\dot{x} = \frac{\partial x_0(I)}{\partial I}\dot{I} = 0 \quad \text{by (2.1)}.$$

Heuristically, we can come to the same conclusion with just a little thought. The invariant manifold is defined by fixing the x variables. Thus if we start a trajectory with this particular fixed value of the x variables for its initial condition (respecting the domain of I in the definition of $\mathcal{M}_0$), then this trajectory leaves the manifold only if this fixed value of x changes. However, by definition, this does not happen since $x_0(I)$ is a fixed point of the x component of (2.1), and so $\dot{x} = 0$.

Since $\mathcal{M}_0$ is an invariant manifold we can consider the restriction to $\mathcal{M}_0$ of the unperturbed normal form (2.1). The resulting vector field is given by

$$\dot{I} = 0,$$
$$\dot{\theta} = \Omega(x_0(I), I),$$

with solutions given by

$$I = I_0 = \text{constant}, \qquad \theta(t) \equiv \Omega(x_0(I), I)t + \theta_0.$$

Next we turn to the term *normal hyperbolicity*. This generalises the notion of a hyperbolic equilibrium point or periodic orbit to the case where the invariant set is not just a single solution, but rather an invariant manifold of solutions. The term "normal" means that the directions normal to the invariant manifold are hyperbolic (more precisely, exponentially growing and decaying under the linearised dynamics). In our case, the directions normal to $\mathcal{M}_0$ are represented by the coordinate x, and since $x_0(I)$ is a hyperbolic fixed point for the x component of (2.1) exponential growth and decay normal to $\mathcal{M}_0$ is evident. The key feature of normally hyperbolic invariant manifolds is that they persist under perturbations, but we shall return to that later. We stress here that our discussion is heuristic. For precise definitions we refer the reader to Wiggins [19].

Now $\mathcal{M}_0$ has $m+l+1$-dimensional stable and unstable manifolds, which we shall denote by $W^s(\mathcal{M}_0)$ and $W^u(\mathcal{M}_0)$ respectively. There are several ways we can understand this statement. First, we consider the stable manifold of $\mathcal{M}_0$, $W^s(\mathcal{M}_0)$.

The stable manifold is the set of points that approach $\mathcal{M}_0$ as $t \to \infty$ (once we have a handle on this set we have to argue that it is indeed a manifold). Here is a very heuristic way of thinking of it that should enable the reader to arrive at the dimension. At any point of the $m + l$ dimensional manifold $\mathcal{M}_0$ there is one exponentially attracting direction normal to the manifold. Hence, as one considers this exponentially attracting direction at all points of $\mathcal{M}_0$ one realises the $m+l+1$-dimensional stable manifold (where we still have not made the argument for the manifold structure). Similar reasoning can be carried out for the unstable manifold of $\mathcal{M}_0$. Now we make this argument more precise and analytical.

As we have outlined above, it can be easily verified that a solution of (2.1) is given by

$$z^h(t, I, \theta_0) : \begin{cases} x^h(t, I), \\ I, \\ \theta(t, I, \theta_0) = \int_0^t \Omega(x^h(s, I), I)\, ds + \theta_0. \end{cases} \tag{2.3}$$

Indeed, we shall argue that it is a solution that lies in $W^s(\mathcal{M}_0)$ and also in $W^u(\mathcal{M}_0)$. This can easily be seen as follows. In the limit as $t \to \pm\infty$ the x component of (2.3) approaches $x_0(I)$. In other words, from (2.2), the trajectory approaches $\mathcal{M}_0$. Next we consider the dimension of the stable and unstable manifolds from this point of view. Consider the trajectory (2.3) at $t = 0$. The initial condition for that trajectory is given by

$$\left(x^h(0, I), I, \theta_0 \right).$$

Now I and θ_0 are m and l dimensional respectively but, viewing I as fixed, $x^h(0, I)$ is constrained to lie on the trajectory $x^h(t, I)$. Adding up the dimensions, we obtain the result that the stable and unstable manifolds are $m+l+1$ dimensional. We have actually obtained a bit more information. That is, the stable and unstable manifolds intersect in an $m+l+1$ dimensional *homoclinic manifold*.

We want to use (2.3) to get a more explicit parametrisation of $W^s(\mathcal{M}_0) \cap W^u(\mathcal{M}_0)$, and, in particular, deal a bit more analytically with the constraint that $x^h(0, I)$ lies on $x^h(t, I)$. To this end, since (2.1) is autonomous, any time translated solution is also a solution (Wiggins [17], Proposition 1.1.11). Hence, for any $t_0 \in \mathbf{R}$,

$$z^h(t - t_0, I, \theta_0) : \begin{cases} x^h(t - t_0, I), \\ I, \\ \theta(t - t_0, I, \theta_0) = \int_0^t \Omega(x^h(s - t_0, I), I)\, ds + \theta_0, \end{cases} \tag{2.4}$$

is also a solution with initial condition

$$\left(x^h(-t_0, I), I, \theta_0 \right).$$

Therefore the map

$$\begin{pmatrix} t_0 \\ I \\ \theta_0 \end{pmatrix} \mapsto \begin{pmatrix} x^h(-t_0, I) \\ I \\ \theta(-t_0, I, \theta_0) = \theta_0 \end{pmatrix}$$

is a parametrisation of $W^s(\mathcal{M}_0) \cap W^u(\mathcal{M}_0)$ and t_0 can be viewed as the unique "time of flight" to the reference point $x^h(0, I)$. Counting up the parameters, t_0, I, θ_0, we see that there are $1 + m + l$ of them, which is the dimension of $W^s(\mathcal{M}_0) \cap W^u(\mathcal{M}_0)$.

Following this latter statement, we can make several other salient remarks concerning invariant manifolds. Invariant manifolds are surfaces that are "knitted together" from trajectories. From this point of view it is very natural to consider them as being parametrised by the initial conditions of those trajectories, and that is precisely what we have done above. Moreover, from this remark comes also the manifold structure. From basic results on the structure of solutions of ordinary differential equations, if the vector field is C^r, for any r, or analytic, then the solutions depend on the initial conditions in a C^r or analytic manner. Hence our invariant manifolds are just as differentiable as the vector field (2.1). We illustrate the geometry of the invariant manifold structure in Fig. 11.

2.2.1 Fibres: foliations of $W^s(\mathcal{M}_0)$ and $W^u(\mathcal{M}_0)$

We saw above that the $m + l + 1$-dimensional $W^s(\mathcal{M}_0)$ and $W^u(\mathcal{M}_0)$ can be viewed as a union of solutions of (2.1). Indeed, that view is at the heart of

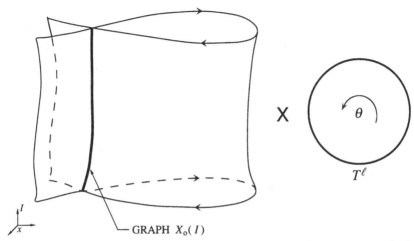

Figure 11: Geometry of the invariant manifold structure for the unperturbed normal form.

what we mean by the term invariant manifold. We say that the trajectories *foliate* the invariant manifold and the individual trajectories are referred to as *fibres* of the foliation. However, there is an alternative foliation of invariant manifolds that may be more useful in certain applications. This is a foliation by surfaces of initial conditions lying on trajectories that approach each other at the "fastest rate". Before explaining this statement in more detail, we will first compute the fibres in $W^u(\mathcal{M}_0)$ by determining the nature of the points that approach each other at the "fastest rate". Let

$$p_j = (t_0^j, I^j, \theta_0^j) \in W^u(\mathcal{M}_0), \qquad j = 1, 2$$

denote two points in $W^u(\mathcal{M}_0)$ through the parametrisation developed above. The trajectories through these points at $t = 0$ are given by

$$z^h(t - t_0^j, I^j, \theta_0^j) : \begin{cases} x^h(t - t_0^j, I^j), \\ I^j, \\ \theta(t - t_0^j, I^j, \theta_0^j) = \int_0^t \Omega(x^h(s - t_0^j, I^j), I^j)\, ds + \theta_0^j, \end{cases} \quad j = 1, 2.$$

We require these trajectories to approach each other as $t \to -\infty$ (remember, these are trajectories in the *unstable* manifold). We examine the implications of this requirement on each component of the trajectories.

Since I is a constant, for the I component we must have

$$I^1 = I^2.$$

Carrying this forward, for the x component we have

$$x^h(t - t_0^1, I^1) - x^h(t - t_0^2, I^2) \to 0 \quad \text{as} \quad t \to -\infty,$$

for $I^1 = I^2$ since $x^h(t - t_0^1, I^1) \to x_0(I^1)$ as $t \to -\infty$. Next we turn to the θ component, which is a bit more tricky. We require

$$\theta(t - t_0^1, I^1, \theta_0^1) - \theta(t - t_0^2, I^1, \theta_0^2)$$

$$= \int_0^t \Omega(x^h(s - t_0^1, I^1), I^1)\, ds - \int_0^t \Omega(x^h(s - t_0^2, I^1), I^1)\, ds + \theta_0^1 - \theta_0^2 \to 0$$

as $t \to -\infty$, where we have used $I^1 = I^2$. In the integrals we change variables by letting $\tau = s - t_0^j$, and after some algebra the requirement that the difference in the θ components approaches zero as $t \to -\infty$ becomes

$$\theta_0^2 = \theta_0^1 + \int_{-t_0^1}^{-t_0^2} \Omega(x^h(\tau, I^1), I^1)\, d\tau.$$

Since the point p_2 was arbitrary we drop the subscript "2" and arrive at the following expression for the *fibre through the point* p_1:

$$I = I^1, \qquad \theta_0 = \theta_0^1 + \int_{-t_0^1}^{-t_0} \Omega(x^h(\tau, I^1), I^1)\, d\tau.$$

Now several remarks are in order.

1. The fibre through p_1 is a one-dimensional curve. This can be seen by the fact that it is given by $m + l$ constraints among the $m + l + 1$ parameters describing points in $W^u(\mathcal{M}_0)$.

2. Recall our earlier discussion of the notion of *normal hyperbolicity* of $\mathcal{M}_0$. In the coordinates of (2.1) the x variables are the "normal variables" while the $I - \theta$ variables describe points on $\mathcal{M}_0$. In more classical terms, the x variables are "fast" variables and the $I - \theta$ variables are "slow" variables. Indeed, normal hyperbolicity is a manifestation of a dichotomy of time scales. In constructing the fibre through an arbitrary point p_1 the fast variables played very little role. Rather, the fibres were constructed through an appropriate initialisation of the slow variables.

3. The fibres of the unstable manifold are referred to as *unstable fibres*. Similarly, the fibres of the stable manifold are referred to as *stable fibres*. Concerning the dimension of the fibres, the unstable (resp. stable) fibres have the same dimension as the number of fast expanding (resp. contracting) directions.

2.2.2   The basepoint of the fibre: intersection with $\mathcal{M}_0$

The unstable fibres intersect $\mathcal{M}_0$ in a unique point called the *basepoint of the unstable fibre*. We compute this point for the unstable fibre through p_1 that we constructed above. We will denote these basepoints by a superscript "b". Clearly we will have

$$I^1 = I^b,$$

since I is a constant. So in this particular normal form it is only the θ component of the fibre that requires some work, and we next turn to this.

We consider the θ component of the trajectory of (2.1) given in (2.4) through p_1 and let $t \to -\infty$. The trajectory approaches $\mathcal{M}_0$, although it does not generally limit to a particular point on $\mathcal{M}_0$. Indeed, upon examination of the θ component of (2.4) in the limit as $t \to -\infty$ one doubts whether this limit exists. For the situation of $\Omega(x_0(I), I) \neq 0$, it does not. The trajectories on $\mathcal{M}_0$ are given by

$$
\begin{aligned}
I(t) &= \text{constant}, \\
\theta(t) &= \Omega(x_0(I), I)t + \theta_0,
\end{aligned}
$$

and trajectories in $W^u(\mathcal{M}_0)$ approach these trajectories in the limit $t \to -\infty$. Hence, it is not surprising that in the limit as $t \to -\infty$ the θ component of (2.4) does not exist since on $\mathcal{M}_0$, $\theta(t)$ is increasing. Motivated by this fact, we compute the limit as $t \to -\infty$ of the θ component of (2.4) *with the asymptotic motion on* $\mathcal{M}_0$ *subtracted away*, and require this difference to be zero:

$$
\lim_{t \to -\infty} \theta(t - t_0^1, I^b, \theta_0^1) - \left(\Omega(x_0(I^b), I^b)t + \theta_0^b \right) = 0.
$$

This computation gives

$$
\theta_0^b = \int_{-t_0^1}^{-\infty} \left(\Omega(x^h(\zeta, I^b), I_b) - \Omega(x_0(I^b), I^b) \right) d\zeta + \theta_0^1.
$$

Hence the basepoint of the unstable fibre through p_1 is given by

$$
z_b : I^1 = I^b, \qquad \theta_0^b = \int_{-t_0^1}^{-\infty} \left(\Omega(x^h(\zeta, I^b), I_b) - \Omega(x_0(I^b), I^b) \right) d\zeta + \theta_0^1.
$$

The unstable (as well as stable) fibres can be parametrised by their basepoints.

So here is the situation developed so far. We have the $m + l$-dimensional normally hyperbolic invariant manifold $\mathcal{M}_0$, with $m+l+1$-dimensional stable and unstable manifolds. The stable and unstable manifolds are foliated by one dimensional curves (fibres) that are parameterised by their intersections with $\mathcal{M}_0$ (their basepoints). The fibres have the property that trajectories with initial conditions on the same fibre approach each other in the appropriate asymptotic direction in time. However, these fibres have an additional important property that we now point out.

2.2.3 The "tracking property" of the fibres

One of the most useful properties of the fibres is that points in a fibre asymptotically approach the same trajectory in $\mathcal{M}_0$. In particular, they approach the trajectory on $\mathcal{M}_0$ with initial condition given by the basepoint of the fibre. This could practically be seen in our derivation of the basepoint of the unstable fibres. Here again, we really only need concern ourselves with the θ component of the trajectories in $W^u(\mathcal{M}_0)$, which, through the point p_1, is given by

$$\theta(t - t_0^1, \theta_0^1, I^b) = \int_{-t_0^1}^{t - t_0^1} \Omega(x^h(\zeta, I^b), I^b) \, d\zeta + \theta_0^1.$$

From the expression for the basepoint of the unstable fibre through p_1 we obtain

$$\theta_0^1 = - \int_{-t_0^1}^{-\infty} \Omega(x^h(\zeta, I^b), I^b) - \Omega(x_0(I^b), I^b) \, d\zeta + \theta_0^b,$$

which, after substituting for the initial condition in the expression for the trajectory in $W^u(\mathcal{M}_0)$, gives

$$\lim_{t \to -\infty} \theta(t - t_0^1, I^b, \theta_0^1) - \left(\Omega(x_0(I^b), I^b)t + \theta_0^b \right) = 0,$$

which proves our claim.

If we denote the unstable fibre with basepoint z_b by $\mathcal{F}^u(z_b)$, then $W^u(\mathcal{M}_0)$ can be represented as a union of these fibres:

$$W^u(\mathcal{M}_0) = \bigcup_{z_b \in \mathcal{M}_0} \mathcal{F}^u(z_b).$$

A similar statement holds for $W^s(\mathcal{M}_0)$, with the obvious modifications. When one considers the perturbed vector field this point of view provides many advantages, as we will see.

2.3 The perturbed system

Now we turn our attention to the perturbed system. The geometrical picture that we developed will form the framework on which we understand the dynamics of the perturbed system. In the phase space of the unperturbed system (2.1) there were three basic geometric structures that we considered:

- The normally hyperbolic invariant manifold $\mathcal{M}_0$.

- The stable and unstable manifolds of $\mathcal{M}_0$, $W^s(\mathcal{M}_0)$ and $W^u(\mathcal{M}_0)$.

- The foliation of $W^s(\mathcal{M}_0)$ and $W^u(\mathcal{M}_0)$ by 1-dimensional stable and unstable fibres, respectively, denoted $\mathcal{F}^s(z_b)$ and $\mathcal{F}^u(z_b)$, $z_b \in \mathcal{M}_0$.

Now we ask the question "what happens to these geometric structures for ϵ small, but not zero?". This question is answered with the persistence theory for normally hyperbolic invariant manifolds and foliations (see Wiggins [19]), which says that for ϵ sufficiently small each of these manifolds persist in the perturbed vector field (2.1). We will denote these perturbed geometric structures by $\mathcal{M}_\epsilon$, $W^s(\mathcal{M}_\epsilon)$, $W^u(\mathcal{M}_\epsilon)$, $\mathcal{F}_\epsilon^s(z_b)$ and $\mathcal{F}_\epsilon^u(z_b)$, $z_b \in \mathcal{M}_\epsilon$. Moreover, each is C^r ϵ-close to their unperturbed counterparts (where r is the degree of differentiability of the perturbed vector field (2.1)).

The reader may feel a bit uneasy with these statements. After all, we did not really "do" anything. We merely proclaimed the persistence of these geometric structures under perturbation. But there is an important point to be made here, and in it lies a key aspect of the wide-ranging utility of the persistence theory for normally hyperbolic invariant manifolds and foliations. Namely, the specific analytical form of the dynamical system is almost irrelevant to the application of the theory (provided it is at least C^r, $r \geq 1$). We have added the term "almost" because one must first identify an invariant manifold for whatever one views as the unperturbed system (hence the efficacy of our unperturbed normal form (2.1)), and then examine its stability properties under the linearised dynamics to determine if it is normally hyperbolic.

Thus the basic "skeleton" of the geometric structure of the unperturbed system (2.1) persists. However, it is important to understand the significance of the term "normal" in the phrase normal hyperbolicity. Even though these manifolds persist for ϵ sufficiently small, the dynamics *on* the manifolds may be very different than that for $\epsilon = 0$.

Dynamics on $\mathcal{M}_\epsilon$. Since $\mathcal{M}_0$ persists as a C^r normally hyperbolic *invariant* manifold for the perturbed problem we can consider the restriction of the perturbed vector field to $\mathcal{M}_\epsilon$, which we denote by

$$\begin{aligned} \dot{I} &= \mathcal{O}(\epsilon), \\ \dot{\theta} &= \Omega(x_0(I), I) + \mathcal{O}(\epsilon). \end{aligned}$$

The $\mathcal{O}(\epsilon)$ terms could be computed perturbatively with a little work. But they are not important for us in the present discussion. Rather, the important feature is that since $\mathcal{M}_\epsilon$ and $\mathcal{M}_0$ are ϵ-close, it follows immediately that the respective restricted vector fields are also ϵ-close. Moreover, the perturbed restricted vector field (2.5) is of a form that has received much attention in the dynamics community. For non-Hamiltonian perturbations it is of the form where the methods of averaging can be applied to study the dynamics and for Hamiltonian perturbations it is of the form where KAM theory and Nekhoroshev's Theorem can be used to study the dynamics (see Arnol'd, Kozlov & Neishtadt [1] and Lochak & Meunier [5] for a comprehensive review and development of these techniques). We will not go into more detail on this now, but rather treat some specific examples in detail and comment on some more general issues afterwards.

Foliations of invariant manifolds. We now give a trivial example that illustrates the key issues and properties associated with foliations of stable

and unstable manifolds of normally hyperbolic invariant manifolds. Consider
the following vector field in the plane:

$$\dot{x} = -\epsilon x,$$
$$\dot{y} = -y.$$

The trajectory through a point (x_0, y_0) is given by

$$x(t) = x_0 e^{-\epsilon t},$$
$$y(t) = y_0 e^{-t}.$$

As might be expected, the unperturbed system is that given when $\epsilon = 0$. In
this case we see that the x-axis is an invariant manifold and the rest of the
plane can be viewed as the stable manifold of this particular normally hyperbolic invariant manifold. For $\epsilon = 0$ all points on the x-axis are equilibrium
points and all trajectories with initial conditions not on the x-axis are vertical
lines (i.e. $x =$ constant) that approach the x-axis as $t \to \infty$. In this case the
fibres of the stable manifold are also vertical lines (see Fig. 12).

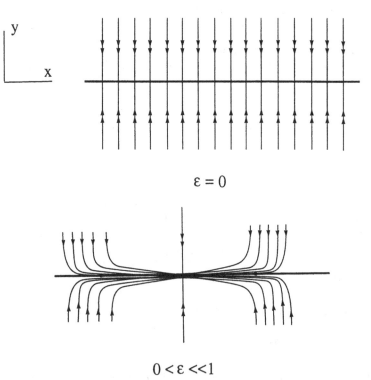

$$\epsilon = 0$$

$$0 < \epsilon \ll 1$$

Figure 12: Fibres and trajectories for the simple example.

Now for $0 < \epsilon \ll 1$ some aspects of the situation change dramatically, others do not. In this case the x-axis is still a normally hyperbolic invariant manifold with the rest of the plane as its stable manifold. However, the dynamics on the x-axis is very different for $\epsilon \neq 0$. In this case the only equilibrium point is the origin and other trajectories may move an $\mathcal{O}(1)$ distance from their unperturbed counterparts for $t = \mathcal{O}(\frac{1}{\epsilon})$ (see Fig. 12). Now what about the stable fibres? Consider two points $p_1 \equiv (x_1, y_1)$ and $p_2 \equiv (x_2, y_2)$ that are not on the x-axis. The difference in the x and y components of the trajectories through these points is given by

$$
\begin{aligned}
x(t, x_1) - x(t, x_1) &= (x_1 - x_2)e^{-\epsilon t}, \\
y(t, y_1) - y(t, y_2) &= (y_1 - y_2)e^{-t}.
\end{aligned}
$$

If we seek the set of initial conditions with the property that the trajectories through them approach each other at the fastest rate then from this expression we see that this requirement is met if we require $x_1 = x_2$. Hence, as in the unperturbed case, the fibres are vertical lines. Alternatively, we could construct the fibres by invoking the "tracking property" described above. Let p_2 be the basepoint of the desired fibre, i.e. $y_2 = 0$, and regard p_1 as an arbitrary point in the stable manifold. Then we want the trajectory through p_1 to approach the trajectory through p_2 as $t \to \infty$. Again, we are led to the requirement of $x_1 = x_2$.

The important feature to learn from this example is that *trajectories* in the stable manifold of the perturbed and unperturbed problems do *not* remain $\mathcal{O}(\epsilon)$ close for infinite time intervals. Indeed, they may drift apart by a large amount. This is a reflection of the fact that the dynamics in the perturbed and unperturbed invariant manifold (the x-axis in this example) can be very different. However, in some applications it is important to relate trajectories in the stable (or unstable) manifolds to certain trajectories in the invariant manifold. Such problems can often be studied very effectively with the fibres, since they do perturb in a C^r ϵ-close manner. In this way one has control over sets of initial conditions that asymptote to the same trajectories on the invariant manifold.

3   Higher Dimensional Melnikov Methods

Next we want to determine if $W^s(\mathcal{M}_\epsilon)$ and $W^u(\mathcal{M}_\epsilon)$ intersect. For $\epsilon = 0$ these two $m + l + 1$-dimensional manifolds coincide along an $m + l + 1$-dimensional manifold in the $m + l + 2$-dimensional phase space (i.e., they are *codimension one* manifolds). This is analogous to the situation of two curves coinciding in the plane in the sense that we should not expect the situation to persist under perturbations. The key issue here is the notion of *transversality*. Two manifolds are said to intersect transversally at a point if the vector space sum of the tangent spaces of the two manifolds at the point span the tangent space of the ambient space at the point. If the manifolds are codimension one, then transversality implies that their one dimensional normal vectors do not coincide (which is not true for $W^s(\mathcal{M}_0)$ and $W^u(\mathcal{M}_0)$). The importance of transversal intersections is that they persist under perturbation. Hence, it is not at all clear that $W^s(\mathcal{M}_\epsilon)$ and $W^u(\mathcal{M}_\epsilon)$ intersect. However, if they do, then there will exist orbits homoclinic to $\mathcal{M}_\epsilon$, which may imply the type of chaotic dynamics discussed earlier.

The intersection of $W^s(\mathcal{M}_\epsilon)$ and $W^u(\mathcal{M}_\epsilon)$ is studied with a higher dimensional version of the *Melnikov function*. Essentially this is the leading order term in a Taylor expansion in ϵ for the distance between $W^s(\mathcal{M}_\epsilon)$ and $W^u(\mathcal{M}_\epsilon)$. Below we outline the key features in its derivation.

Step 1: Set up a moving *homoclinic coordinate system* **for the unperturbed system.** Any point

$$p \in W^s(\mathcal{M}_0) \cap W^u(\mathcal{M}_0)$$

can be described through the parametrisation $p = (x^h(-t_0, I), I, \theta_0)$, developed earlier. At p the vector

$$\nabla \left(H(x^h(-t_0, I), I) - H(x_0(I), I) \right) \equiv N_p,$$

is perpendicular to $W^s(\mathcal{M}_0) \cap W^u(\mathcal{M}_0)$. We define the set of vectors

$$\hat{I} \equiv (\hat{I}_1, \hat{I}_2, \ldots, \hat{I}_m),$$

where the $\hat{I}_i$ are unit vectors in the I_i directions. Then

$$\Pi_p \equiv \text{span}\{N_p, \hat{I}\},$$

is an $m + 1$ dimensional plane defined at each $p \in W^s(\mathcal{M}_0) \cap W^u(\mathcal{M}_0)$. This is what we mean by a *moving homoclinic coordinate system*. At each $p \in W^s(\mathcal{M}_0) \cap W^u(\mathcal{M}_0)$, these manifolds intersect Π_p *transversely* in the same m-dimensional intersection set (see Fig. 13).

Step 2: The perturbed structure with respect to the homoclinic coordinates. Using the following:

- persistence of normally hyperbolic invariant manifolds,

- persistence of transversality,

- regular perturbation theory,

we can conclude that at each $p \in W^s(\mathcal{M}_0) \cap W^u(\mathcal{M}_0)$, N_p intersects $W^s(\mathcal{M}_\epsilon)$ in a point p_ϵ^s, and $W^u(\mathcal{M}_\epsilon)$ in a point p_ϵ^u (see Fig. 14).

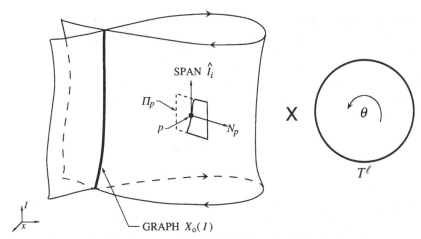

Figure 13: The intersection of $W^s(\mathcal{M}_0)$ and $W^u(\mathcal{M}_0)$ with Π_p.

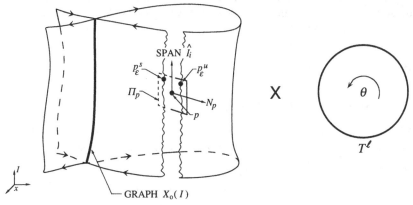

Figure 14: The intersection of $W^s(\mathcal{M}_\epsilon)$ and $W^u(\mathcal{M}_\epsilon)$ with N_p.

Step 3: Use steps 1 and 2 to develop a measure of the distance between $W^s(\mathcal{M}_\epsilon)$ and $W^u(\mathcal{M}_\epsilon)$ at the point p. The distance between $W^s(\mathcal{M}_\epsilon)$ and $W^u(\mathcal{M}_\epsilon)$ at the point p, in the context of the above set-up, is given by

$$
\begin{aligned}
d(p, \epsilon) &= \frac{\langle p_\epsilon^u - p_\epsilon^s, N_p \rangle}{\|N_p\|} \\
&= \epsilon \frac{\langle \frac{\partial p_\epsilon^u}{\partial \epsilon}\big|_{\epsilon=0} - \frac{\partial p_\epsilon^s}{\partial \epsilon}\big|_{\epsilon=0}, N_p \rangle}{\|N_p\|} + O(\epsilon^2),
\end{aligned}
$$

where $\langle \cdot, \cdot \rangle$ denotes the standard Euclidean inner product.

Step 4: Apply Melnikov's trick to develop a computable expression for the leading order term. Using the linear variational equation about the homoclinic manifold, Melnikov derived a linear ordinary differential equation whose solution, evaluated at $t = 0$, is the numerator in the leading order term of the distance measurement described above. This expression is termed *the Melnikov function*. The same "trick" works in this higher dimensional setting and gives the following:

$$\left\langle \left.\frac{\partial p_\epsilon^u}{\partial \epsilon}\right|_{\epsilon=0} - \left.\frac{\partial p_\epsilon^s}{\partial \epsilon}\right|_{\epsilon=0}, N_p \right\rangle \equiv M(t_0, I, \theta_0; \mu)$$

$$= \int_{-\infty}^{\infty} \langle N_p, (g^x, g^I, g^\theta) \rangle (z^h(t - t_0, I, \theta_0)) \, dt.$$

Nondegenerate zeroes of this function correspond to transverse intersections of $W^s(\mathcal{M}_\epsilon)$ and $W^u(\mathcal{M}_\epsilon)$ that are $\mathcal{O}(\epsilon)$ close to the point p. Hence, in this way we can detect orbits homoclinic to $\mathcal{M}_\epsilon$. However, $\mathcal{M}_\epsilon$ need not be an equilibrium point, periodic orbit, or quasiperiodic orbit. Therefore to understand the dynamical consequences of these homoclinic orbits we need to understand the dynamics on $\mathcal{M}_\epsilon$. We will address this particular issue through a series of examples.

4 Examples

4.1 Feedback control systems

We consider the example of a nonlinear oscillator subjected to weak feedback control as embodied in the following class of feedback control systems:

$$\ddot{x}_1 + \delta \dot{x}_1 + k(x_1)x_1 = -I + F(t),$$
$$\dot{I} + \epsilon I = \epsilon \gamma \left(x_1 - x_r(t) \right),$$

where the parameters and terms in this equation have the following meaning:

δ is the damping;

$k(x_1)$ is the nonlinear spring constant;

$\frac{1}{\epsilon}$ is the time constant;

γ is the gain;

$F(t)$ is the external force;

$x_r(t)$ is the desired position history.

We ask the following two questions:

1. If the nonlinear oscillator is chaotic, can feedback be introduced to *control* the chaos?

2. Can feedback introduce chaos into the system?

In answering these questions we need two things. One is to understand mechanisms that can give rise to chaos (e.g. homoclinic orbits), and the second is to have analytical techniques for determining where these mechanisms occur in phase space, as well as parameter space. These are exactly the ideas that we have been developing in this article. To illustrate them we consider an explicit example of this class of feedback control systems given by

$$
\begin{aligned}
\dot{x}_1 &= x_2, \\
\dot{x}_2 &= x_1 - x_1^3 - I - \epsilon\delta x_2, \\
\dot{I} &= \epsilon(\gamma x_1 - \alpha I + \beta\cos\theta), \\
\dot{\theta} &= 1,
\end{aligned}
$$

which has a 4 dimensional phase space with coordinates $(x_1, x_2, I, \theta) \in \mathbf{R} \times \mathbf{R} \times \mathbf{R} \times T^1$. Note that this is a special case of (2.1) with $m = l = 1$. Complete details of this system can be found in Wiggins [16].

Phase space structure of the unperturbed system. Setting $\epsilon = 0$ gives

$$
\begin{aligned}
\dot{x}_1 &= x_2, \\
\dot{x}_2 &= x_1 - x_1^3 - I, \\
\dot{I} &= 0, \\
\dot{\theta} &= 1.
\end{aligned}
$$

The $x_1 - x_2$ component is Hamiltonian with Hamiltonian function given by

$$
H(x_1, x_2; I) = \frac{x_2^2}{2} - \frac{x_1^2}{2} + \frac{x_1^4}{4} + x_1 I.
$$

It can be verified that for $-\frac{2}{3\sqrt{3}} < I < \frac{2}{3\sqrt{3}}$ the $x_1 - x_2$ component has a hyperbolic fixed point $x_0(I)$ connected by two homoclinic orbits (the $x_1 - x_2$ component is the equation of motion for a particle in a double-well potential). The phase space structure appears as in Fig. 15.

The Poincaré map. Next we analyse the system for $\epsilon \neq 0$. As a preliminary step we will use the periodicity of the vector field to cast the problem in the form of a three-dimensional Poincaré map.

A cross-section (or "Poincaré section") to the phase space is given by

$$
\Sigma = \{(x_1, x_2, I, \theta) \,|\, \theta = 0\},
$$

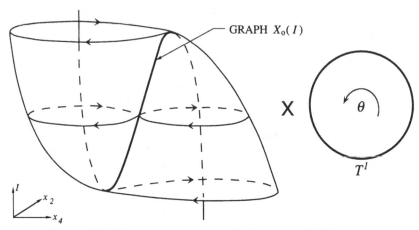

Figure 15: Phase space structure of the unperturbed system.

and the map of this cross-section into itself is given by

$$P : \Sigma \mapsto \Sigma,$$
$$(x_1(0), x_2(0), I(0)) \mapsto (x_1(2\pi), x_2(2\pi), I(2\pi)).$$

In the context of the Poincaré map (and the three-dimensional Poincaré section), for $\epsilon = 0$ the graph of $x_0(I)$ is a one-dimensional normally hyperbolic invariant manifold whose two-dimensional stable and unstable manifolds coincide.

The perturbed structure: dynamics on the invariant manifold. For $\epsilon \neq 0$ but small, these basic structures persist, although the two-dimensional stable and unstable manifolds no longer need to intersect. However, for the moment, we focus our attention on the dynamics on the one-dimensional normally hyperbolic invariant manifold. This manifold has the form

$$x_\epsilon(I, \theta) = x_0(I) + \mathcal{O}(\epsilon).$$

Restricting the vector field to this invariant manifold gives

$$\dot{I} = \epsilon (\gamma x_{0,1}(I) - \alpha I + \cos \theta) + \mathcal{O}(\epsilon^2),$$
$$\dot{\theta} = 1,$$

where $x_0(I) \equiv (x_{0,1}(I), x_{0,2}(I))$. This vector field is in the appropriate form for an immediate application of the method of averaging (i.e. the ϵ is multiplying the leading order term in the I component of the equation).

The Averaging Theorem tells us that hyperbolic fixed points of the $\mathcal{O}(\epsilon)$ averaged part of the I component of this vector field, i.e.

$$\dot{I} = \frac{\epsilon}{2\pi} \int_0^{2\pi} [\gamma x_{1,0}(I) - \alpha I + \beta \cos \theta] \, d\theta$$

$$= \epsilon \left[\gamma x_{1,0}(I) - \alpha I \right], \tag{4.1}$$

correspond to hyperbolic periodic orbits in the full system. In turn, these correspond to hyperbolic fixed points of the Poincaré map.

It can be shown that there exist at most three fixed points of (4.1) at

$$I = 0, \quad \pm \frac{\gamma}{\alpha} \sqrt{1 - \frac{\gamma}{\alpha}}.$$

So for $\frac{\gamma}{\alpha} < 1$ there are three fixed points and for $\frac{\gamma}{\alpha} > 1$ there is one fixed point. These fixed points have the following stability characteristics.

$\boxed{\gamma > \alpha}$

Hyperbolic fixed point at $(x_1, x_2, I) = (0, 0, 0)$ having a two dimensional unstable manifold and a one dimensional stable manifold.

$\boxed{\frac{2\alpha}{3} < \gamma < \alpha}$

Three hyperbolic fixed points at $(x_1, x_2, I) = \left(\pm \sqrt{1 - \frac{\gamma}{\alpha}}, 0, \pm \frac{\gamma}{\alpha} \sqrt{1 - \frac{\gamma}{\alpha}} \right)$ and $(0, 0, 0)$, where $(0, 0, 0)$ has a two dimensional stable manifold and a one dimensional unstable manifold, and $\left(\pm \sqrt{1 - \frac{\gamma}{\alpha}}, 0, \pm \frac{\gamma}{\alpha} \sqrt{1 - \frac{\gamma}{\alpha}} \right)$ have two dimensional unstable manifolds and one dimensional stable manifolds.

$\boxed{\gamma < \frac{2\alpha}{3}}$

Hyperbolic fixed point at $(x_1, x_2, I) = (0, 0, 0)$ having a two dimensional stable manifold and a one dimensional unstable manifold.

In Fig. 16 we illustrate these fixed points and their local stable and unstable manifolds.

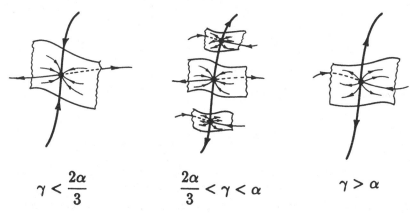

$$\gamma < \frac{2\alpha}{3} \qquad\qquad \frac{2\alpha}{3} < \gamma < \alpha \qquad\qquad \gamma > \alpha$$

Figure 16: The fixed points of the Poincaré map and their local stable and unstable manifolds.

The perturbed structure: intersection of the stable and unstable manifolds of the hyperbolic fixed points of the Poincaré map. The fixed points on the perturbed normally hyperbolic invariant manifold are parametrised by I. To determine if the stable and unstable manifolds of a particular fixed point intersect we compute the Melnikov function with the I value of its arguments fixed at that particular value. The appropriate Melnikov functions are given by

$I = 0$:

$$M^{\pm}(t_0, \alpha, \beta, \gamma, \delta) = -\frac{4\delta}{3} + 4\gamma \pm \sqrt{2}\beta\pi \operatorname{sech}\frac{\pi}{2}\cos t_0, \qquad (4.2)$$

$I = -\frac{\gamma}{\alpha}\sqrt{1 - \gamma/\alpha}$:

$$
\begin{aligned}
M_l^{\pm}(t_0, \alpha, \beta, \gamma, \delta) &= -4\delta\left[\frac{d}{3} + \frac{\gamma b}{\sqrt{2}\alpha}\left(\frac{\pi}{2} \pm \sin^{-1}\sqrt{\frac{2\alpha}{\gamma}}b\right)\right] \\
&\quad + 2\gamma\left[2d - 2\sqrt{2}b\left(\frac{\pi}{2} \pm \sin^{-1}\sqrt{\frac{2\alpha}{\gamma}}b\right)\right] \\
&\quad \mp 2\sqrt{2}\pi\beta\frac{\sinh\left(\frac{1}{d}\sin^{-1}\sqrt{\frac{\alpha}{\gamma}}d\right)}{\sinh\frac{\pi}{d}}\cos t_0, \qquad (4.3)
\end{aligned}
$$

$I = +\frac{\gamma}{\alpha}\sqrt{1 - \gamma/\alpha}$:

$$
\begin{aligned}
M_u^{\pm}(t_0, \alpha, \beta, \gamma, \delta) &= -4\delta\left[\frac{d}{3} + \frac{\gamma b}{\sqrt{2}\alpha}\left(\frac{\pi}{2} \pm \sin^{-1}\sqrt{\frac{2\alpha}{\gamma}}b\right)\right] \\
&\quad + 2\gamma\left[2d - \sqrt{2}b\left(\frac{\pi}{2} \pm \sin^{-1}\sqrt{\frac{2\alpha}{\gamma}}b\right)\right] \\
&\quad \pm 2\sqrt{2}\pi\beta\frac{\sinh\left(\frac{1}{d}\sin^{-1}\sqrt{\frac{\alpha}{\gamma}}d\right)}{\sinh\frac{\pi}{d}}\cos t_0, \qquad (4.4)
\end{aligned}
$$

where, on the $I = \pm\frac{\gamma}{\alpha}\sqrt{1 - \gamma/\alpha}$ levels, the superscript "+" refers to the larger homoclinic loop, and where

$$b = \sqrt{1 - \frac{\gamma}{\alpha}}, \qquad d = \sqrt{\frac{3\gamma}{\alpha} - 2}.$$

We present graphs of $(4.2)^{\pm}$, $(4.4)_u^{\pm}$, and $(4.3)_l^{\pm}$ in Fig. 17 for $\alpha = 1, \beta = 1$. In the region bounded by $(4.4)_u^-$ and $(4.3)_l^-$, the stable and unstable manifolds of the hyperbolic fixed point on the $I = \pm\frac{\gamma}{\alpha}\sqrt{1 - \gamma/\alpha}$ levels corresponding to the small homoclinic orbit intersect transversely; in the region bounded by $(4.4)_u^+$ and $(4.3)_l^+$, the stable and unstable manifolds of the hyperbolic fixed

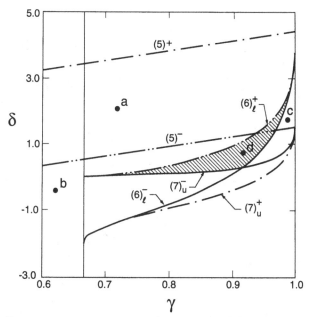

Figure 17: Parameter regions corresponding to zeros of the Melnikov functions (with $\alpha = \beta = 1$).

point on the $I = \pm \frac{\gamma}{\alpha}\sqrt{1 - \gamma/\alpha}$ levels corresponding to the larger homoclinic orbit intersect transversely; and in the region bounded by $(4.2)^+$ and $(4.2)^-$, both branches of the stable and unstable manifolds of the hyperbolic fixed point on the $I = 0$ level intersect transversely. In Fig. 18, we illustrate the behaviour of the stable and unstable manifolds of the Poincaré map for the four different parameter values indicated in Fig. 17.

4.1.1   Control of chaos

Let us consider the situation of the gain γ going to zero. In this case, the $x_1 - x_2$ component of the vector field decouples from the I component. Thus, I can be solved for as an explicit function of time which is asymptotically periodic ($I \sim \epsilon\beta\sin t + \mathcal{O}(\epsilon^2)$ as $t \to \infty$), and the solution can be substituted into the $x_1 - x_2$ components of the equation. The result is an equation for a periodically forced Duffing oscillator

$$\begin{aligned}
\dot{x}_1 &= x_2 \\
\dot{x}_2 &= x_1 - x_1^3 - \epsilon[\delta x_2 + \beta\sin t] + \mathcal{O}(\epsilon^2).
\end{aligned}$$

The original Melnikov method [6] gives a curve in $\beta - \delta$ space above which transverse homoclinic orbits to a hyperbolic periodic orbit exist. This curve

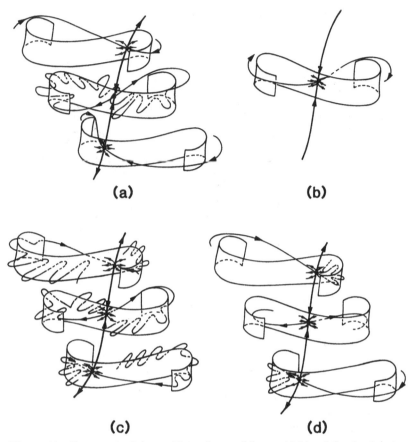

Figure 18: Geometry of the stable and unstable manifolds of the fixed points of the Poincaré map in the different regions of the parameter space as indicated in Fig. 17.

is given by

$$\beta = \frac{4\delta}{3\sqrt{2\pi}}\cosh(\pi/2).$$

From (4.2) we see that a similar curve, above which there exist transverse homoclinic orbits to a hyperbolic periodic orbit on the $I = 0$ level in the presence of nonzero gain γ, is given by

$$\beta = \frac{\left(\frac{4\delta}{3} - 4\gamma\right)}{\sqrt{2\pi}}\cosh\frac{\pi}{2}. \tag{4.5}$$

These curves are shown in Fig. 19. Thus, we see from Fig. 19 that the effect of the gain is to lower the boundary, and hence increase the area of the region in $\beta - \delta$ space in which Smale horseshoes are present in the dynamics.

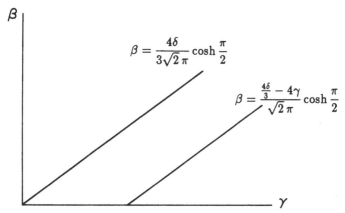

Figure 19: The shift in the boundary for the existence of Smale horseshoes in the $\beta - \delta$ parameter space as a result of feedback.

4.1.2 Introduction of chaos via feedback

The fact that the feedback loop has introduced chaos into the system is evident from Fig. 17. The fixed points and horseshoes on the $I = \pm\frac{\gamma}{\alpha}\sqrt{1 - \gamma/\alpha}$ levels are there solely as a result of the feedback.

Hence, we conclude that feedback control can alter the boundary of chaos (in parameter space) of a system *but* feedback control may introduce *new* chaotic parameter regimes into the system.

4.2 Quasiperiodically forced Duffing oscillator

There has been much work done on chaos in *periodically* forced one-degree-of-freedom oscillators. In this case there is a canonical reduction to a two-dimensional Poincaré map. However, if the forcing is quasiperiodic with two or more incommensurate frequencies, then reduction to a two-dimensional Poincaré map is not possible. Nevertheless, a similar framework applies if we include the phases of each of the frequency components as new dependent variables. We illustrate this in the following example with two frequencies.

We consider the Duffing oscillator subjected to two-frequency quasiperiodic forcing which can be written in the following form

$$
\begin{aligned}
\dot{x}_1 &= x_2, \\
\dot{x}_2 &= -\epsilon\gamma x_2 + \frac{1}{2}(1 - x_1^2)x_1 + \epsilon f_1 \cos\theta_1 + \epsilon f_2 \cos\theta_2, \\
\dot{\theta}_1 &= \omega_1, \\
\dot{\theta}_2 &= \omega_2,
\end{aligned}
$$

which has a four-dimensional phase space with coordinates $(x_1, x_2, \theta_1, \theta_2) \in \mathbf{R} \times \mathbf{R} \times T^1 \times T^1$.

For ω_1 and ω_2 incommensurate (i.e. $\frac{\omega_1}{\omega_2}$ is an irrational number) the usual two-dimensional Poincaré map construction does not go through.

A Poincaré map.   However, we can construct a *three*-dimensional Poincaré map as follows. Define the cross-section to the vector field

$$\Sigma = \{(x_1, x_2, \theta_1, \theta_2) \,|\, \theta_2 = 0\}.$$

Then the Poincaré map of this cross-section into itself is given by

$$P : \Sigma \mapsto \Sigma,$$
$$(x_1(0), x_2(0), \theta_{10}) \mapsto \left(x_1\left(\frac{2\pi}{\omega_2}\right), x_2\left(\frac{2\pi}{\omega_2}\right), \theta_{10} + 2\pi\frac{\omega_1}{\omega_2} \right).$$

Phase space structure for the Poincaré map for $\epsilon = 0$. P has a one-dimensional torus T_0 given by

$$T_0 = \{(x_1, x_2, \theta_1) \in \Sigma \,|\, x_1 = x_2 = 0\}$$

which has two-dimensional stable and unstable manifolds that coincide (this is nothing more than than cartesian product of the phase space of the un-damped, undriven Duffing equation with a circle). We illustrate the phase space structure in Fig. 20.

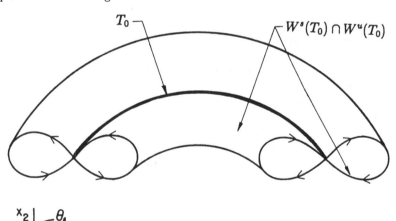

Figure 20: The normally hyperbolic invariant one-torus and its coinciding stable and unstable manifolds for the unperturbed Poincaré map (cut-away one half view).

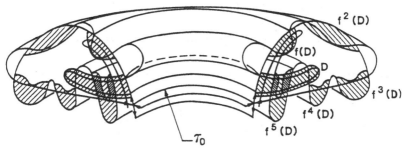

Figure 21: The geometry associated with the transverse intersection of the stable and unstable manifolds of a normally hyperbolic invariant torus and the analogue of the horseshoe construction.

Phase space structure for the Poincaré map for $\epsilon \neq 0$. For $\epsilon \neq 0$ and small, T_0, along with its stable and unstable manifolds, persists. We next want to determine if these stable and unstable manifolds intersect. This will imply the existence of orbits homoclinic to a normally hyperbolic invariant torus.

4.2.1 Mechanisms for chaos in quasiperiodically forced oscillators

Earlier we looked at the case of orbits homoclinic to equilibrium points and periodic orbits as mechanisms for generating chaos in the sense of the shift dynamics. In quasiperiodically forced oscillators there is a different mechanism: orbits homoclinic to normally hyperbolic invariant tori. Such orbits can also generate the shift dynamics. References for this are Wiggins [16] (where appropriate versions of the Conley-Moser conditions are developed), Scheurle [10], Meyer & Sell [7], Stoffer [14, 15], Beigie, Leonard & Wiggins [3] and Yagasaki [20]. In Fig. 21 we illustrate the situation of intersecting stable and unstable manifolds of a normally hyperbolic invariant torus and the analogue of the horseshoe construction. Unlike the situation of the transverse intersection of the stable and unstable manifolds of a hyperbolic periodic point, the stable and unstable manifolds of a normally hyperbolic invariant torus can intersect transversely in higher dimensional sets than a point. This leads to many new geometric properties. Examples can be found in Beigie, Leonard & Wiggins [3] and Wiggins [18].

The quasiperiodic Melnikov function: orbits homoclinic to tori. The Melnikov function in this case is given by

$$M(t_0, \theta_{10}; \gamma, f_1, f_2, \omega_1, \omega_2) = -\frac{2\sqrt{2}}{3}\gamma \pm 2\pi f_1 \omega_1 \mathrm{sech}\frac{\pi\omega_1}{\sqrt{2}} \sin(\omega_1 t_0 + \theta_{10})$$
$$\pm 2\pi f_2 \omega_2 \mathrm{sech}\frac{\pi\omega_2}{\sqrt{2}} \sin \omega_2 t_0.$$

The zeroes of the quasiperiodic Melnikov function are given by

$$\gamma = \pm\frac{3\pi}{\sqrt{2}}f_1\omega_1\text{sech}\frac{\pi\omega_1}{\sqrt{2}}\sin(\omega_1 t_0 + \theta_{10}) \pm \frac{3\pi}{\sqrt{2}}f_2\omega_2\text{sech}\frac{\pi\omega_2}{\sqrt{2}}\sin\omega_2 t_0.$$

Hence, in the 5-dimensional parameter space $(\gamma, f_1, f_2, \omega_1, \omega_2)$ the following 4-dimensional surface

$$\gamma = \frac{3\pi}{\sqrt{2}}f_1\omega_1\text{sech}\frac{\pi\omega_1}{\sqrt{2}} + \frac{3\pi}{\sqrt{2}}f_2\omega_2\text{sech}\frac{\pi\omega_2}{\sqrt{2}} \qquad (4.6)$$

is the bifurcation surface describing the bifurcation to orbits homoclinic to the invariant 1-torus of the 3-dimensional Poincaré map.

Obviously, visualisation of this 4-dimensional surface in 5 dimensions poses some difficulties. However, there is some special structure in this problem that ameliorates the difficulties slightly. We will present bifurcation *curves* in the $\omega_1 - \omega_2$ plane for fixed values of γ, f_1 and f_2. We do this because of the fact that, since (4.6) is *linear* in γ, f_1 and f_2 (this is the special structure), it can be shown that $\gamma - f_1 - f_2$ space is divided into regions separated by planes (with the planes independent of ω_1 and ω_2) and the regions define topologically distinct bifurcation curves in the $\omega_1 - \omega_2$ plane. Details of this analysis can be found in Ide & Wiggins [4].

In Fig. 22 the $\gamma - f_1 - f_2$ space is shown divided into five regions by a collection of planes. The five regions are denoted R^0, R^1, R_1^2, R_2^2 and R^3 (the subscript i on R_i^2, $i = 1, 2$, indicates that $f_i > f_j$, $j = 1, 2$, $j \neq i$). These

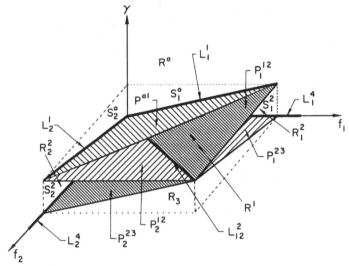

Figure 22: The regions in $\gamma - f_1 - f_2$ space that give rise to qualitatively distinct homoclinic bifurcation curves in the $\omega_1 - \omega_2$ space.

regions are separated by the planes P^{01}, P_1^{12}, P_1^{23}, P_2^{12} and P_2^{23}. The latter four of these planes meet at the line L_{12}^2 whose f_1 and f_2 coordinates satisfy $f_1 = f_2$. The plane P^4 is specified by $\gamma = 0$.

In the $\gamma - f_1$ plane we denote the boundary of R_0 by S_1^0, P^{01} and P_1^{12} meet in the line L_1^1, P_1^{23} and P^4 meet in the line L_1^4, and S_1^2 is the region between L_1^1 and L_1^4. A similar notation holds in the $\gamma - f_2$ plane.

The bifurcation curves in the $\omega_1 - \omega_2$ plane corresponding to parameter values in these different regions are shown in Fig. 23 where ω_m denotes the unique maximum of the function $\omega \operatorname{sech} \frac{\pi \omega}{\sqrt{2}}$.

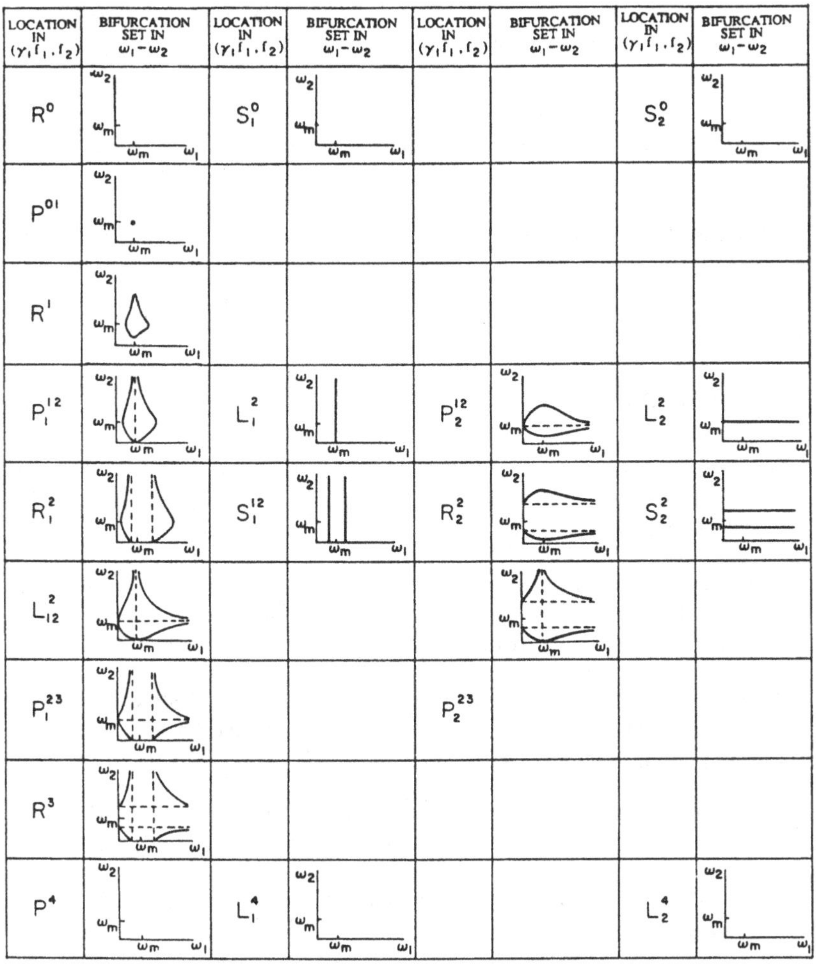

LOCATION IN $(\gamma_1 f_1, f_2)$	BIFURCATION SET IN $\omega_1 - \omega_2$	LOCATION IN $(\gamma_1 f_1, f_2)$	BIFURCATION SET IN $\omega_1 - \omega_2$	LOCATION IN $(\gamma_1 f_1, f_2)$	BIFURCATION SET IN $\omega_1 - \omega_2$	LOCATION IN $(\gamma_1 f_1, f_2)$	BIFURCATION SET IN $\omega_1 - \omega_2$
R^0		S_1^0				S_2^0	
P^{01}							
R^1							
P_1^{12}		L_1^2		P_2^{12}		L_2^2	
R_1^2		S_1^{12}		R_2^2		S_2^2	
L_{12}^2							
P_1^{23}				P_2^{23}			
R^3							
P^4		L_1^4				L_2^4	

Figure 23: Homoclinic bifurcation curves in the $\omega_1 - \omega_2$ space corresponding to the parameter values in the different regions shown in Fig. 22.

- For some values of (γ, f_1, f_2) (e.g. in R^1) and for a given value of ω_1, there may be no values of ω_2 such that homoclinic orbits occur. However, for some values of ω_1, there may be a window of ω_2 values in which homoclinic orbits occur.

- For some values of (γ, f_1, f_2) (e.g. in R_1^2), and for any value of ω_2, there is a window of ω_1 values in which homoclinic orbits occur. Outside of this window no homoclinic orbits occur.

- For some values of (γ, f_1, f_2) (e.g. in R^3), and for some value of ω_2, there is a window of ω_1 values in which homoclinic orbits occur. Outside of this window no homoclinic orbits occur. For other values of ω_2, homoclinic orbits occur for all ω_1 values.

Hence, chaos may be created or destroyed in periodically excited systems by the introduction of additional excitation frequencies.

Acknowledgements

I am grateful to Carl Bird for typing an early draft of these lectures.

References

[1] Arnol'd, V.I., Kozlov, V.V. and Neishtadt, A.I., *Dynamical Systems III*. Encyclopedia of Mathematical Sciences Series, ed. V.I. Arnol'd, Springer-Verlag, New York, 1988.

[2] Arnol'd, V.I., Afrajmovich, V.S., Il'yashenko, Y.S. and Sil'nikov, L.P., Bifurcation Theory, *Dynamical Systems V*, Encyclopedia of Mathematical Sciences, ed. V.I. Arnol'd, Springer-Verlag, New York, 1994.

[3] Beigie, D., Leonard, A. and Wiggins, S., Chaotic transport in the homoclinic and heteroclinic tangle regions of quasiperiodically forced two dimensional dynamical systems, *Nonlinearity* **4**, 775-819, 1991.

[4] Ide, K. and Wiggins, S., The bifurcation to homoclinic tori in the quasiperiodically forced duffing oscillator, *Physica D* **34**, 169-182, 1989.

[5] Lochak, P. and Meunier, C., *Multiphase Averaging for Classical Systems*, Springer-Verlag, New York, 1988.

[6] Melnikov, V.K., On the stability of the center for time periodic perturbations, *Trans. Moscow Math. Soc.* **12**, 1-57, 1963.

[7] Meyer, K.R., Sell, G.R., Melnikov transforms, Bernoulli bundles, and almost periodic perturbations, *Trans. Am. Math. Soc.* **314**, 63-105, 1989.

[8] Palis, J. and deMelo, W., *Geometric Theory of Dynamical Systems: An Introduction*, Springer-Verlag, New York, 1982.

[9] Palis, J. and Takens, F., *Hyperbolicity and Sensitive Chaotic Dynamics at Homoclinic Bifurcations*, Cambridge University Press, Cambridge, 1993.

[10] Scheurle, J., Chaotic solutions of systems with almost periodic forcing, *J. Appl. Math. Phys. (ZAMP)* **37**, 12-26, 1986.

[11] Sil'nikov, L.P., A case of the existence of a denumerable set of periodic motions, *Sov. Math. Dokl.* **6**, 163-166, 1965.

[12] Smale, S., *The Mathematics of Time. Essays on Dynamical Systems, Economic Processes and Related Topics*, Springer-Verlag, New York, 1980.

[13] Sparrow, C., *The Lorenz Equations*, Springer-Verlag, New York, 1982.

[14] Stoffer, D., Transversal homoclinic points and hyperbolic sets for non-autonomous maps I, *J. Appl. Math. Phys. (ZAMP)* **39**, 518-549, 1988.

[15] Stoffer, D., Transversal homoclinic points and hyperbolic sets for non-autonomous maps II. *J. Appl. Math. Phys. (ZAMP)* **39**, 783-812, 1988.

[16] Wiggins, S., *Global Bifurcations and Chaos: Analytical Methods*, Springer-Verlag, New York, 1988.

[17] Wiggins, S., *Introduction to Applied Nonlinear Dynamical Systems and Chaos*, Springer-Verlag, New York, 1990.

[18] Wiggins, S., *Chaotic Transport in Dynamical Systems*, Springer-Verlag Interdisciplinary Applied Mathematical Sciences Series, 1992.

[19] Wiggins, S., *Normally Hyperbolic Invariant Manifolds in Dynamical Systems*, Springer-Verlag, New York, 1994.

[20] Yagasaki, K., Chaotic dynamics of quasiperiodically forced oscillators detected by Melnikov's method. *SIAM J. Math. Anal.* **23**, 1230-1254, 1992.

Some Recent Developments in Nonlinear Elasticity and its Applications to Materials Science

J.M. Ball
Department of Mathematics,
Heriot-Watt University,
Edinburgh

1 Introduction

Vital to a proper understanding of mathematical models of nature is knowledge of the possible singularities possessed by solutions of the governing equations together with physical interpretations of these singularities. The underlying theme of this article will be to discuss this within the context of elasticity theory, for which such knowledge has important implications for the understanding of the behaviour of materials. In elasticity theory materials are characterised by their corresponding free-energy functions, and so we need to know how different assumptions on the free-energy function can give rise to various types of singularity in the solution. This philosophy might lead one also towards a strategy for proving that solutions for certain partial differential equations (PDE's) are smooth, namely to classify the possible singularities of a class of such equations and to identify hypotheses on the form of the equations to eliminate each kind of singularity in turn. This is *not* how regularity theory for PDE's is generally approached, but we will see some hints that such a strategy might be viable in the future.

We consider three main topics:

(a) 'Mathematically well-behaved materials', i.e. the study of a special class of free-energy functions leading to the existence of energy minimisers with mathematically desirable properties, and including some commonly used models of particular materials.

(b) Fracture and its mathematical description.

(c) Materials that can undergo phase transformations.

We shall see that the description of (b) and (c) cannot be subsumed under (a) because of the nature of the corresponding singularities.

2   Elasticity and Energy Minimisation

We begin by reviewing how to describe the deformation of an elastic body. We consider only the static case, that is, only deformations independent of time are considered, and no attempt is made to describe the dynamical process by which these deformations arise. This is partly for simplicity, but largely because our understanding of dynamics is, by comparison, very limited. Let $\Omega \subset \mathbf{R}^n$ represent the reference configuration of an elastic body which we use to label its material points. Using the Lagrangian description, the position of a point $x \in \Omega$ in a typical deformed configuration is denoted by $y(x) \in \mathbf{R}^n$ (see Fig. 1). Thus $y : \Omega \to \mathbf{R}^n$. Of course, the cases of interest are $n = 1, 2$ or 3. The gradient of y at x is written $Dy(x)$, and can be identified with the $n \times n$ matrix of partial derivatives $\left(\frac{\partial y_i}{\partial x_j}(x) \right)$.

The material will be assumed homogeneous, so that its mechanical response (i.e. the stress corresponding to a given strain) is independent of the point x. Note that this is more restrictive than saying that Ω is occupied by the same material at each point, due to the possibility of pre-existing stresses.

The first question we consider is concerned with the choice of functions which should be used for a suitable mathematical model of deformations.

To be physically acceptable a deformation y should be *invertible*. This is essentially to avoid interpenetration of matter. However we still might want to allow cases where self-contact occurs. For example the deformation of a bar illustrated in Fig. 2 is an acceptable deformation, although it is not invertible on $\bar{\Omega}$. Also cases where y is invertible in Ω except on a set of measure zero could be considered (e.g. when the inverse image of a point is a line or a surface). In order that the deformation y be orientation preserving and have the same orientation as in the reference configuration we require that

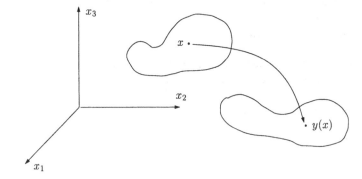

Figure 1: Description of deformation of a three-dimensional elastic body.

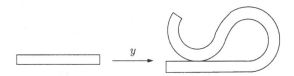

Figure 2: A deformation which is invertible on Ω but not on $\bar{\Omega}$.

$$\det Dy(x) > 0 \tag{2.1}$$

almost everywhere (a.e.) in Ω. If $y \in C^1(\Omega; \mathbf{R}^n)$, (2.1) implies local invertibility, this being a consequence of the Inverse Function Theorem. However local invertibility does not imply global invertibility as the deformation of a bar illustrated in Fig. 3 shows.

If $y \notin C^1(\bar{\Omega}; \mathbf{R}^n)$ then (2.1) does not even imply local invertibility. As an example consider the mapping $y : (r, \theta) \mapsto (r, 2\theta)$ of the unit disc in $\mathbf{R}^2$ to itself. It is easily checked that the gradient is bounded a.e. and that its determinant is a positive constant. However y is not invertible in any neighbourhood of the origin. A key tool for proving global, and even local, invertibility is *degree theory*. For various results see [10, 26, 33, 44, 65, 68].

Suppose now that Ω is a bounded, open connected set with a Lipschitz boundary $\partial\Omega$ (i.e. each point $x \in \partial\Omega$ has a neighbourhood in which $\partial\Omega$ can be represented as the graph of a Lipschitz function and in which Ω lies on one side of $\partial\Omega$). Let $\partial\Omega_1 \subset \partial\Omega$ be a portion of the boundary with $\mathcal{H}^{n-1}(\partial\Omega_1) > 0$, where $\mathcal{H}^{n-1}$ denotes $(n-1)$–dimensional Hausdorff measure (i.e. surface area). We consider the boundary condition

$$y|_{\partial\Omega_1} = f \tag{2.2}$$

for a given mapping $f : \partial\Omega_1 \to \mathbf{R}$. We impose no boundary conditions on the remaining part of the boundary $\partial\Omega\backslash\partial\Omega_1$. For the variational problem we

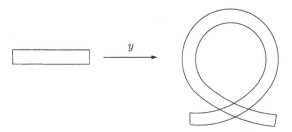

Figure 3: A locally invertible deformation that is not globally invertible.

consider this will formally correspond to requiring that the applied traction vanish on $\partial\Omega\backslash\partial\Omega_1$. The elastic energy corresponding to the deformation y is defined as

$$I(y) = \int_\Omega W(Dy) \, dx, \qquad (2.3)$$

where $W : M^{n\times n} \to [0,\infty]$ is the free-energy (or stored-energy) function and $M^{m\times n}$ denotes the space of real $m \times n$ matrices. The assumption that $W \geq 0$ is made simply for convenience (it is natural to assume that W is bounded below, and adding a constant to W does not change the problem). We assume that $W(A)$ is finite and continuous for $\det A > 0$, and that

$$W(A) \to \infty \qquad \text{as} \qquad \det A \to 0+, \qquad (2.4)$$

which physically means that an infinite amount of energy is required to crush the material down to zero volume. Consistent with the constraint (2.1) we suppose that $W(A) = \infty$ for $\det A \leq 0$. Thus W is continuous with respect to the natural topology on $[0,\infty] = [0,\infty) \cup \{\infty\}$.

Question. *Does there exist a $\bar{y}$ minimising I subject to the boundary condition (2.2)?*

The above question is imprecise since the class of admissible functions is not clearly specified. Also it is not clear what we mean by Dy. These issues can be explored by considering the underlying physics of the problem. In general enlarging the set of admissible functions can change the nature of the solution. It may result in the infimum of the energy functional changing (the *Lavrentiev phenomenon*) and affect whether or not the infimum is attained. It is thus important to assess whether or not a given set of admissible functions has a physical justification. For example the absolute minimum of the energy

$$I(y) = \int_0^1 y'^2 \, dx$$

subject to the boundary condition $y(0) = 0$, $y(1) = 1$ is attained by the well-known Cantor function when y' is interpreted as the limit of a difference quotient, with minimum value zero. However such a function would usually be regarded as too irregular to represent an acceptable physical deformation. Among more regular functions, such as we use below, the minimum value is 1, attained by the function $y(x) = x$. Other more surprising examples can be found in [16]. In our case we make the set of admissible functions precise by using the standard notion of Sobolev spaces (see, for example, [2, 25, 32, 52, 54]).

For $1 \leq p \leq \infty$ define the Sobolev space $W^{1,p} = W^{1,p}(\Omega; \mathbf{R}^n)$ to be the set of equivalence classes of mappings $y : \Omega \to \mathbf{R}^n$ which together with their first order weak derivatives belong to L^p. This means that

$$\|y\|_{1,p} = \left(\int_\Omega (|\,y\,|^p + |\,Dy\,|^p) \, dx \right)^{\frac{1}{p}} < \infty \qquad \text{if } 1 \leq p < \infty,$$

$$\|y\|_{1,\infty} \;=\; \operatorname*{ess\,sup}_{x \in \Omega}(\mid y(x) \mid + \mid Dy(x) \mid) < \infty,$$

where $Dy = \left(\frac{\partial y_i}{\partial x_j}\right)$ denotes the matrix of first order weak partial derivatives of y, defined to satisfy

$$\int_\Omega \frac{\partial y_i}{\partial x_j}\varphi\,dx = -\int_\Omega y_i\frac{\partial\varphi}{\partial x_j}\,dx \qquad \text{for all} \ \varphi \in C_0^\infty(\Omega). \qquad (2.5)$$

In (2.5) $C_0^\infty(\Omega)$ denotes the space of infinitely differentiable real-valued functions on Ω which vanish outside a compact subset of Ω. We can now pose our question more precisely by setting

$$\mathcal{A} = \left\{y \in W^{1,1} : y|_{\partial\Omega_1} = f, \quad I(y) < \infty\right\}. \qquad (2.6)$$

In (2.6) the boundary condition is to be understood in the sense of *trace* (for the precise meaning see the general references cited above).

Question (precise formulation). *Assume that $\mathcal{A}$ is not empty. Does there exist a function $\bar{y} \in \mathcal{A}$ minimising I ?*

3 Mathematically Well-Behaved Materials

In this section we consider a class of materials for which a positive answer to the above question can be given, and highlight some limitations of the existing theory.

3.1 Statement of the existence theorem

Suppose $n = 3$ and that

(H1) W is *polyconvex* i.e. there exists a convex function $g : M^{3\times3} \times M^{3\times3} \times (0,\infty) \to \mathbf{R}$ such that $W(A) = g(A, \operatorname{cof} A, \det A)$ for all $A \in M^{3\times3}$ with $\det A > 0$;

(H2) $W(A) \geq c_0(\mid A \mid^p + \mid \operatorname{cof} A \mid^q) - c_1$, where $c_0 > 0$, $p \geq 2$, $q \geq \frac{3}{2}$.

Here $\operatorname{cof} A$ denotes the matrix of cofactors of A. Then we have the following result.

Theorem 1 *Let* (H1) *and* (H2) *hold. Then there exists $\bar{y}$ minimising I in $\mathcal{A}$. Furthermore $\det D\bar{y}(x) > 0$ for a.e. $x \in \Omega$.*

3.2 Commentary

Theorem 1 is of the type first proved in Ball [8] under somewhat stronger growth hypotheses. It was refined by Ball & Murat [17], who assumed that $q \geq \frac{p}{p-1}$, and then by Müller, Qi & Yan [48] in the version stated here.

The theorem avoids the blunder of assuming W to be convex. This is not a physically realistic assumption. For example, a consequence of convexity of W would be that any equilibrium configuration (solution of the Euler-Lagrange equation) is an absolute minimiser, in particular ruling out buckling phenomena. Also note that the set $M_+^{3\times 3} = \{A \in M^{3\times 3} : \det A > 0\}$ is not convex, so that W cannot both be convex and satisfy (2.4).

To give examples of materials satisfying (H1), (H2) and (2.4), let us first look at some general properties of the free-energy function. First, W should satisfy the frame-indifference condition

$$W(A) = W(RA) \quad \text{for all } A \in M_+^{3\times 3}, \ R \in SO(3).$$

This condition reflects the fact that rigid rotations of a body do not change its energy, and is assumed to hold for all materials.

The symmetry of the material is expressed by the condition

$$W(A) = W(AR) \quad \text{for all } A \in M_+^{3\times 3}, \ R \in \mathcal{S},$$

where we will assume that the symmetry group $\mathcal{S}$ of the material is a subgroup of $SO(3)$. A particular case is when the material has no preferred grain. This corresponds to the case $\mathcal{S} = SO(3)$, and the material is called *isotropic*. Using the polar decomposition theorem it can be shown that W is frame-indifferent and isotropic if and only if W has the representation

$$W(A) = \Phi(v_1, v_2, v_3)$$

for $\det A > 0$, where v_1, v_2 and v_3 are the singular values of A (i.e. the eigenvalues of $(A^T A)^{\frac{1}{2}}$) and Φ is symmetric with respect to permutations of the v_i. Writing

$$
\begin{aligned}
\phi(\alpha) &= v_1^\alpha + v_2^\alpha + v_3^\alpha - 3, \\
\psi(\beta) &= (v_2 v_3)^\beta + (v_3 v_1)^\beta + (v_1 v_2)^\beta - 3,
\end{aligned}
$$

then the well-known Ogden materials ([56, 57]) have the form

$$\Phi(v_1, v_2, v_3) = \sum_{i=1}^{M} a_i \phi(\alpha_i) + \sum_{i=1}^{N} b_i \psi(\beta_i) + h(v_1 v_2 v_3),$$

where $a_i > 0$, $b_i > 0$, $\alpha_1 \geq \alpha_2 \geq \ldots \geq \alpha_M \geq 1$, $\beta_1 \geq \beta_2 \geq \ldots \geq \beta_N \geq 1$, and where $h : (0, \infty) \to [0, \infty)$ is convex, with $h(\delta) \to \infty$ as $\delta \to 0$. These free-energy functions satisfy (H1) and (2.4). If $\alpha_1 \geq 2$ and $\beta_1 \geq \frac{3}{2}$ then they also satisfy (H2). For appropriate values of the constants good fits can be obtained for the behaviour of rubber-like materials.

In order to place the polyconvexity hypothesis (H1) in perspective we consider some related classes of free-energy functions. We say that W is *rank-one convex* if and only if it is convex in every rank-one direction, that is $t \mapsto W(A + t\lambda \otimes \mu)$ is a convex function of t for every $A \in M^{3 \times 3}$ and $\lambda, \mu \in \mathbf{R}^3$. If $t \mapsto W(A + t\lambda \otimes \mu)$ is strictly convex for $t \geq 0$ whenever A, $A + \lambda \otimes \mu \in M_+^{3 \times 3}$ we say that W is *strictly rank-one convex*.

In the case when $W \in C^2(M_+^{3 \times 3})$, rank-one convexity is equivalent to the Legendre-Hadamard condition

$$\frac{d^2}{dt^2} W(A + t\lambda \otimes \mu)|_{t=0} \geq 0 \qquad \text{for all } A \in M_+^{3 \times 3}, \ \lambda, \mu \in \mathbf{R}^3,$$

that is

$$\frac{\partial^2 W}{\partial A_{ij} \partial A_{kl}}(A) \lambda_i \mu_j \lambda_k \mu_l \geq 0 \qquad \text{for all } A \in M_+^{3 \times 3}, \ \lambda, \mu \in \mathbf{R}^3, \qquad (3.1)$$

where we have used the summation convention. If strict inequality holds in (3.1) whenever λ and μ are nonzero, then W is said to be *strongly elliptic*.

We say that W is *quasiconvex* if

$$\int_E W(Dv)\,dx \geq \int_E W(A)\,dx$$

for every bounded open set $E \subset \mathbf{R}^3$ and $v \in C^1(E; \mathbf{R}^3)$ with $v = Ax$ in a neighbourhood of ∂E. It can be shown by a scaling argument that the definition of quasiconvexity is independent of E. The definitions of polyconvexity, rank-one convexity and quasiconvexity extend in the obvious way to arbitrary dimensions, i.e. to integrands $W : M^{m \times n} \to [0, \infty]$. The following implications then hold:

$$\text{polyconvexity} \implies \text{quasiconvexity} \implies \text{rank-one convexity.}$$

In particular all the above notions coincide with the usual convexity when either $n = 1$ or $m = 1$.

The importance of rank-one matrices lies in the fact that a deformation $y \in W^{1,1}(\Omega; \mathbf{R}^m)$ can have its gradient equal to A and B respectively above and below a specific $(n - 1)$–dimensional hyperplane intersecting Ω if and only if the difference $A - B$ is a rank-one matrix. To justify this statement, known as the *Hadamard jump condition*, let y be such that

$$Dy = \begin{cases} A & \text{if } x \cdot \mu > k, \\ B & \text{if } x \cdot \mu < k, \end{cases}$$

for some $k \in \mathbf{R}$ and unit vector $\mu \in \mathbf{R}^n$. Applying the continuity on the interface (the existence of a continuous representative follows from the boundedness of Dy and the Sobolev Embedding Theorem), we obtain $C\mu^\perp = 0$ whenever $\mu^\perp \cdot \mu = 0$, where $C = A - B$. Since

$$
\begin{aligned}
(C - C\mu \otimes \mu)\mu &= 0, \\
(C - C\mu \otimes \mu)\mu^\perp &= 0 \qquad \text{for } \mu^\perp \cdot \mu = 0,
\end{aligned}
$$

it follows that $C = C\mu \otimes \mu$ and therefore $A - B = \lambda \otimes \mu$ for some $\lambda \in \mathbf{R}^m$. Conversely if $A - B = \lambda \otimes \mu$ one can easily find a deformation whose gradient is equal to A and B respectively below and above an $(n-1)$–dimensional hyperplane with normal μ. This construction will be used frequently in Section 5 below.

It would be less restrictive in Theorem 1 to replace the polyconvexity of W by quasiconvexity. There are various results in the calculus of variations of this type following from the pioneering work of C.B. Morrey [45, 46] (see, for example, [1, 43]). However the assumptions made are not consistent with the growth conditions of nonlinear elasticity. Quasiconvexity is a difficult condition to verify as it is not defined in a pointwise manner, and no equivalent pointwise condition (i.e. one depending on W and its derivatives at an arbitrary matrix) is known. For over 40 years it was a conjecture arising out of the work of Morrey that quasiconvexity was equivalent to rank-one convexity. However in 1992 V. Šverák [66] produced a counterexample for a quartic $W = W(Dy)$ for the case when $y : \Omega \subset \mathbf{R}^n \to \mathbf{R}^m$ and $n \geq 2, m \geq 3$. The case when $m = n = 2$ remains open, and there is some evidence that rank-one convexity might imply quasiconvexity in this case (see [27, 28, 59, 60]).

Theorem 1 (and many others similar to it by nature) guarantees the existence of at least one (global) minimiser. However it does not say anything about how smooth such a minimiser is. In particular one is interested to know whether or not the Euler-Lagrange equations are satisfied. To derive the Euler-Lagrange equations we need to calculate the first variation of I. Let $\varphi : \Omega \to \mathbf{R}^3$ be smooth with $\varphi|_{\partial\Omega_1} = 0$. Then $\bar{y} + t\varphi \in \mathcal{A}$ for any $t \in \mathbf{R}$ and so

$$
\frac{d}{dt} I(\bar{y} + t\varphi)|_{t=0} = \int_\Omega D_A W(D\bar{y}) \cdot D\varphi \, dx = 0 \qquad (3.2)
$$

provided this derivative exists and is given by the anticipated expression in (3.2). To justify this we need the existence of the limit

$$
\lim_{t \to 0} \frac{1}{t} [I(\bar{y} + t\varphi) - I(\bar{y})] = \lim_{t \to 0} \int_\Omega \frac{1}{t} [W(D\bar{y} + tD\varphi) - W(D\bar{y})] \, dx, \qquad (3.3)
$$

where we assume that $W \in C^1(M_+^{3\times 3})$ say. But it is not obvious how to pass to the limit inside the integral in (3.3) unless we have additional information on $\bar{y}$, so that for example the integrands are bounded pointwise by an integrable function (for the application of the Dominated Convergence Theorem).

This is the case if $\bar{y} \in W^{1,\infty}$ with $\det D\bar{y} \geq \varepsilon > 0$ for some $\varepsilon > 0$. However, the only readily available piece of information is that $I(\bar{y}) < \infty$. In fact there are one-dimensional counterexamples (see [16]) of minimisers which do not satisfy the corresponding Euler–Lagrange equation, even for elliptic polynomial integrands. The Euler–Lagrange equation is then an elliptic ordinary differential equation; weak solutions of such equations are smooth, but the minimisers are not. It is not known whether the minimiser $\bar{y}$ given by Theorem 1 satisfies (3.2). However, it is possible to derive other weak forms of the Euler-Lagrange equation under supplementary growth hypotheses on W (see [12, 19]).

The condition of *strict rank-one convexity* is intimately related to the possible occurrence of the type of singularity we previously considered in which Dy jumps across a hyperplane. In fact (see [9]), under the very mild supplementary hypothesis that the integrand W has a minimiser, strict rank-one convexity is necessary and sufficient for the nonexistence of such singularities in solutions to the Euler-Lagrange equations.

The general question of whether the minimisers in Theorem 1 are smooth is unresolved. A well-known theorem of Evans [31] implies 'partial regularity' of minimisers $\bar{y}$ of integrals of the form (2.3) with W satisfying a strict form of quasiconvexity, though again unfortunately under growth hypotheses inconsistent with (2.4). More precisely he proves that under his hypotheses $\bar{y}$ is smooth off a closed set S of measure zero. Such a result would be very interesting for the minimisers in Theorem 1, especially if S were nonempty for physically reasonable W. The existence of points $x_0 \in S$ where $|D\bar{y}(x_0)| = \infty$ in some sense might, for example, have something to do with the onset of fracture, though as we discuss in Section 4, the energy functional I needs modification for actual fracture to be described. The counterexample of Nečas [53] shows that if $\bar{y}$ is in fact smooth (so that S is empty) this must have something to do with either the special form of W or with special features of 3 (or low) dimensions. Even under the growth hypotheses of Evans essentially nothing is known about regularity up to the boundary.

For recent advances in the regularity theory of minimisers in elasticity see [18, 19, 20, 21, 34, 35, 64].

In Theorem 1 the minimisation was carried out in a space $\mathcal{A}$ of deformations y satisfying the condition $\det Dy > 0$ a.e.. As we saw in Section 2 such deformations need not even be locally invertible, and to be physically realistic we should instead minimise over globally invertible deformations. In the case of pure displacement boundary conditions (i.e. $\partial\Omega_1 = \partial\Omega$) this turns out not to be a serious issue, since in fact the minimiser $\bar{y}$ in Theorem 1 can be shown to be a homeomorphism provided the boundary data f is consistent with invertibility and provided somewhat stronger growth conditions are imposed on W (see [10, 65]). However, for mixed boundary conditions ($\partial\Omega_1 \neq \partial\Omega$) there is no escaping the need for a new approach, and one way is to proceed as in Ciarlet & Nečas [26]. However, there is much to be done before the true complications of self-contact can be satisfactorily addressed.

4   Fracture and Cavitation

4.1   Towards a theory of fracture

Why and how do materials break? These are the questions studied in the important technological subject of *fracture mechanics*. By far the largest proportion of work in fracture mechanics concerns the possible extension and evolution of *pre-existing cracks* under different loadings. In contrast, comparatively little is said about fracture initiation. Although there are isolated rigorous treatments of parts of classical fracture mechanics (see, for example, [58]), from the mathematical perspective fracture mechanics is in a somewhat primitive state, reflecting the difficulty of considering (or eliminating from consideration on some rational basis) very complicated potential fracture patterns. There is perhaps now the beginnings of a more general and rigorous approach to be seen in the work on problems with 'free discontinuity sets' based on ideas of geometric measure theory (see [3, 4, 30, 51]). But there is a need to link these new ideas with classical work in fracture mechanics, a necessary preliminary to a more comprehensive theory.

A first mathematical difficulty which arises is that we cannot use Sobolev spaces to directly model fracture in general. In the typical fracture patterns illustrated in Fig. 4 the deformations y shown jump across a two-dimensional surface, and hence $y \notin W^{1,1}$.

Let us consider the static case. The idea in [3, 4, 30, 51] is to seek to minimise the modified energy

$$I(y, K) = \int_{\Omega \setminus K} W(Dy)\, dx + k\mathcal{H}^2(K), \qquad k > 0, \qquad (4.1)$$

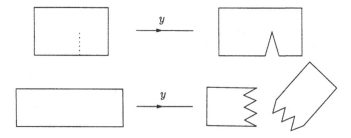

Figure 4: Examples of deformations y involving fracture: such deformations do not belong to Sobolev spaces.

subject to some boundary conditions. Here Ω and W are as before, and the new unknown K is the *crack set*. The second term is the simplest way to take into account the contribution of the surface energy due to the creation of cracks. Here $\mathcal{H}^2(K)$ denotes the two-dimensional Hausdorff measure of K and coincides with usual definitions of surface area for smooth sets K. The competing functions y are supposed to belong to $W^{1,1}(\Omega \backslash K; \mathbf{R}^3)$. This allows them to have a jump across the crack set.

The main difficulty here is that we want to minimise I without making any *a priori* assumptions on the set K. A key technical device introduced by Ambrosio & de Giorgi in [30, 3, 4] is to remove the dependence of the integral in (4.1) on K by using the space $SBV(\Omega; \mathbf{R}^3)$. This space is defined in terms of the space $BV(\Omega; \mathbf{R}^3)$ of mappings $y : \Omega \to \mathbf{R}^3$ of *bounded variation*, consisting of those $y \in L^1(\Omega; \mathbf{R}^3)$ whose gradient Dy is a bounded measure. This gradient can be decomposed as the sum of a part ∇y that is absolutely continuous with respect to 3-dimensional Lebesgue measure, and a singular part $D^s y$:

$$Dy = \nabla y \, dx + D^s y.$$

The singular part can be further decomposed as

$$D^s y = (y^+ - y^-)\nu_y d\mathcal{H}^2|_{S_y} + Cy,$$

where ν_y is the measure theoretic normal to the set of *jump points* S_y, $y^\pm$ denote the traces of y on either side of S_y, and Cy is the *Cantor part* of Dy. $SBV(\Omega; \mathbf{R}^3)$ is now defined as consisting of those $y \in BV(\Omega; \mathbf{R}^3)$ whose Cantor part $Cy = 0$. We can now replace the problem of minimising (4.1) by that of minimising the associated functional

$$I(y) = \int_\Omega W(\nabla y) \, dx + k\mathcal{H}^2(S_y) \tag{4.2}$$

over a suitable subset of $SBV(\Omega; \mathbf{R}^3)$. There is now a considerable literature on existence and regularity of minimisers for functionals of the form (4.2) (see, for example, [5, 6, 24, 29]). However, from the point of view of applications to fracture mechanics a missing ingredient seems to be a calculation of a general variation for I about a given $\bar{y}$ in the direction of a nearby y having a possibly very different set of jump points.

4.2 Cavitation

Cavitation is a common failure mechanism in polymers, such as rubber, characterised by the formation of (roughly spherical) holes (see [36]). Fortunately it is possible to express cavitation in Sobolev spaces with an unmodified energy functional

$$I(y) = \int_\Omega W(Dy) \, dx. \tag{4.3}$$

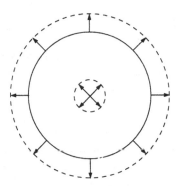

Figure 5: Radial cavitation.

In the simplest situation of radial cavitation, a hole forms at the centre of a ball whose outer boundary is radially displaced (see Fig. 5).

The corresponding deformation y is discontinuous, since it maps a solid ball into one with a hole. For such a deformation to have finite energy (4.3) W must have restricted growth. In fact, if the material is 'strong', namely

$$W(A) \geq C(1+ \mid A \mid^\gamma), \qquad A \in M_+^{3\times3} \tag{4.4}$$

for some $\gamma > 3$, then it follows immediately from the Sobolev Embedding Theorem that any deformation y with $I(y) < \infty$ is continuous. Thus *cavitation type singularities* cannot occur in strong materials.

Let Ω be the open ball $B = \{x \in \mathbf{R}^3 : |x| < 1\}$ and consider radial deformations of the type

$$y(x) = \frac{r(|x|)}{|x|}x. \tag{4.5}$$

Such deformations y map spheres to spheres and displace points of B in the direction of their position vectors with respect to the centre of B. We assume that the material is isotropic. The question is now whether or not there exists any radial minimiser whose corresponding r in (4.5) satisfies $r(0) > 0$. After some calculations it can be shown that the singular values of Dy are

$$v_1 = r'(|x|), \qquad v_2 = v_3 = \frac{r(|x|)}{|x|}.$$

Therefore restricting to the class of radial functions $I(y) = \int_B W(Dy)\,dx$ takes the form

$$E(r) = 4\pi \int_0^1 R^2 \Phi(r', r/R, r/R)\,dR.$$

We can try to minimise E over the set of admissible functions

$$\mathcal{A}_\lambda = \{r \in W^{1,1}(0,1) : r' > 0 \text{ a.e., } r(1) = \lambda, r(0) \geq 0\},$$

where the parameter λ represents the boundary displacement (one can formulate the traction problem in a similar fashion).

Consider now the special free-energy function

$$\Phi(v_1, v_2, v_3) = v_1^\alpha + v_2^\alpha + v_3^\alpha + h(v_1 v_2 v_3), \qquad 1 < \alpha < 3, \qquad (4.6)$$

where h is smooth and satisfies

$$h'' > 0, \qquad \lim_{\delta \to \infty} \frac{h(\delta)}{\delta} = \lim_{\delta \to 0+} h(\delta) = \infty.$$

Note that this free-energy function satisfies (H1) but not (H2) for the given range of α.

It can be shown that for λ less than a critical value λ_{cr} the trivial solution $r(R) = \lambda R$ (i.e. $y(x) = \lambda x$) is the only (global) minimiser of E. However if $\lambda > \lambda_{cr}$ there is an exchange of stability and a nontrivial cavitating solution r_λ with $r_\lambda(0) > 0$ becomes the global minimiser, the trivial solution becoming unstable. The bifurcation diagram Fig. 6 shows this.

The above example has been studied for a wider range of W in [11] (for different methods see [61, 63]). There it is shown that the minimisers r_λ satisfy the Euler-Lagrange equations and hence provide singular weak solutions to the 3D equations of nonlinear elastostatics. However it is not known in general whether the radial deformations still remain minimisers when one considers variations that are not radially symmetric, or whether λ_c gives the correct critical boundary displacement when arbitrary nonradial deformations are allowed. If the growth of $W(A)$ is less than quadratic in A then it has been shown by James & Spector [40] that for certain free-energy functions further reduction of energy is possible by introducing cylindrical voids along radial lines. For a recent review of cavitation in elastic materials see [39].

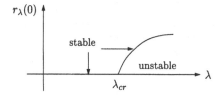

Figure 6: Bifurcation of the radial minimiser r_λ from the trivial solution $r(R) = \lambda R$.

The phenomenon of cavitation is a striking illustration of the fact that the function space chosen is *part of the mathematical model*. If we choose to minimise in, say, $W^{1,p}$ for $p > 3$ then the unique minimiser y of I subject to $y|_{\partial B} = \lambda x$ is $\bar{y}(x) = \lambda x$ (because W given through (4.6) is quasiconvex) and cavitation is not predicted. For general formulations of this kind of minimisation see Giaquinta, Modica & Souček [38] and the simplification of Müller [47]. If we choose to minimise in the smaller spaces considered in [38] (which has the advantage of a general 3D existence theory), the only way to recover cavitation is to suppose that the body already possesses small pre-existing holes. Although such holes do not seem to have been observed, they were postulated in [36] as being the precursors of cavitation. Since there are experiments [55, 37] in which cavitation appears at well-defined and repeatable geometric locations (see [62] for a corresponding theory based on our model above), in order to use the spaces in [38] we would need to assume that the pre-existing holes permeate the body, leading to a reference domain Ω with very complicated geometrical structure. This makes these spaces unattractive for modelling cavitation. Sivaloganathan [61] considered the case when the reference domain contains a pre-existing hole of radius ε, with a zero traction condition on the inner boundary $|x| = \varepsilon$, and showed that the corresponding radial minimisers (which are smooth) converge to r_λ as $\varepsilon \to 0$. This suggests a consistency between the theory in [38] and that based on a straightforward minimisation of I. However, this has not been proved in general, and the limiting case of a domain with infinitely many holes would require a difficult process of homogenisation.

The existence of minimisers of I for W such as (4.6) and arbitrary (non-radial) boundary conditions is an open question. However, Müller & Spector [49] have proved existence when a surface energy term is added to $I(y)$ consisting of a constant multiple of the area of $\partial y(\Omega)$ (a functional related to, but different from, (4.1)).

It is worth contrasting the different effects of a rapid growth of $W(A)$ with A in the theory of cavitation and in that based on (4.1). As we have seen, if W satisfies (4.4) for some $\gamma > 3$ then cavitation is prevented. But for the more realistic model (4.1) rapid growth of W is expected to *promote* fracture for a body under tension, since the body can release a lot of elastic energy by fracturing. By working in a Sobolev space we tacitly assume that the surface energy is infinite, negating this effect. Of course cavitation can presumably occur in the model based on (4.1), but then it has to compete energetically with other types of fracture.

5 Phase Transformations

5.1 A minimisation problem with nonattainment and the formation of microstructure.

To motivate the ideas appearing in this section let us look at the following minimisation problem for the scalar function $u = u(x_1, x_2)$ defined on the square $\Omega = (0,1) \times (0,1)$:

$$\text{Min } I(u) = \int_\Omega [(u_{x_1}^2 - 1)^2 + u_{x_2}^2]\, dx_1\, dx_2 \qquad (5.1)$$

subject to $u(x_1, 0) = 0$.

By constructing a suitable minimising sequence it can be shown that the infimum of I is zero. Roughly speaking one should consider a sequence $\{u^{(j)}\}$ such that the partial derivatives satisfy $u_{x_1}^{(j)} = \pm 1$ and $u_{x_2}^{(j)} = 0$. However in order for the boundary condition to be satisfied a 'transition layer' should exist close to $x_2 = 0$ so that the function matches itself with $u(x_1, 0) = 0$. Putting this in a more precise way consider

$$\bar{u}(x_1, x_2) = \begin{cases} x_1 \phi(x_2) & \text{if } 0 \leq x_1 \leq \frac{1}{2}, \\ (1 - x_1)\phi(x_2) & \text{if } \frac{1}{2} \leq x_1 \leq 1, \end{cases}$$

where $\phi(x_2) = x_2$ if $0 \leq x_2 \leq 1$ and $\phi(x_2) = 1$ if $x_2 \geq 1$. Extending $\bar{u}$ as a one-periodic function of x_1 to $\mathbf{R} \times (0, \infty)$ one can now define

$$u^{(j)}(x_1, x_2) = \frac{1}{j}\bar{u}(jx_1, jx_2).$$

Then $Du^{(j)}(x_1, x_2) = (\bar{u}_{x_1}, \bar{u}_{x_2})(jx_1, jx_2)$ is uniformly bounded and consequently $I(u^{(j)}) \to 0$ as $j \to \infty$. However this infimum is *not attained*. If $I(u) = 0$ for some $u \in W^{1,1}$ with $u(x_1, 0) = 0$ then $u_{x_2} = 0$ and hence, integrating in the x_2 direction, $u = 0$. But this implies $I(u) = 1$, a contradiction.

5.2 Young measures

The above problem is a typical example where the infimum of a functional over a class of admissible functions is not attained, and minimising sequences develop a microstructure in which the gradient oscillates more and more finely. The idea of generalised curves or *Young measures* was first introduced and used by L.C. Young to tackle such problems (see his book [69]). In general a Young measure $(\nu_x)_{x \in \Omega}$ is a family of probability measures that gives the limiting distribution of values of a sequence of functions. The following is one possible characterisation [13]. Let $\Omega \subset \mathbf{R}^n$ be an open set. Suppose

$z^{(j)} : \Omega \to \mathbf{R}^m$ is a sequence of measurable functions. For given x, j and δ define $\nu_{x,\delta}^{(j)}$ to be the probability distribution of the values of $z^{(j)}(p)$ as p is chosen uniformly at random from the ball $B(x; \delta)$. Then

$$\nu_x = \lim_{\delta \to 0} \lim_{\mu \to \infty} \nu_{x,\delta}^{(\mu)},$$

with the limit to be understood in the sense of weak* convergence of probability measures. (In general the $z^{(\mu)}$ correspond to a subsequence of the $z^{(j)}$.) For recent surveys of Young measures see [7, 42, 67].

As an example, the reader can check that the Young measure corresponding to the sequence of gradients $u_{x_1}^{(j)}, u_{x_2}^{(j)}$ of any minimising sequence $u^{(j)}$ for the problem (5.1) is given by

$$\nu_x = \frac{1}{2}\delta_{(1,0)} + \frac{1}{2}\delta_{(-1,0)},$$

where $\delta_{(\pm 1,0)}$ denotes the Dirac mass at $(\pm 1, 0)$.

5.3 Microstructure arising from a phase transformation

Now to see what happens in the case of elastic crystals let us look at the problem of a phase transformation in a single crystal of a material, such as the binary alloy Indium-Thallium, where the high temperature phase (*austenite*) has cubic symmetry and the low temperature phase (*martensite*) has tetragonal symmetry. We suppose that θ_c is the temperature at which the phase transformation occurs, and take the reference configuration to be the cubic phase at the temperature $\theta = \theta_c$. We use nonlinear elasticity with free-energy function $W = W(Dy, \theta)$ depending on the temperature. The phase transformation is described by an exchange of stability. Denoting by $K(\theta)$ the set of minimisers of $W(\cdot, \theta)$ in $M^{3 \times 3}$ we assume that

$$\begin{aligned}
K(\theta) &= SO(3) & \text{for } \theta > \theta_c, \\
K(\theta) &= \textstyle\bigcup_{i=1}^{3} SO(3)U_i & \text{for } \theta < \theta_c, \\
K(\theta_c) &= SO(3) \cup \textstyle\bigcup_{i=1}^{3} SO(3)U_i,
\end{aligned}$$

where

$$\begin{aligned}
U_1 &= \mathrm{diag}\,(\eta_2, \eta_1, \eta_1), \\
U_2 &= \mathrm{diag}\,(\eta_1, \eta_2, \eta_1), \\
U_3 &= \mathrm{diag}\,(\eta_1, \eta_1, \eta_2).
\end{aligned}$$

The *lattice parameters* η_i in general will depend on temperature. Note that these assumptions are consistent with the frame-indifference and cubic symmetry of W. We call the connected components of $K(\theta)$ *energy wells*; thus,

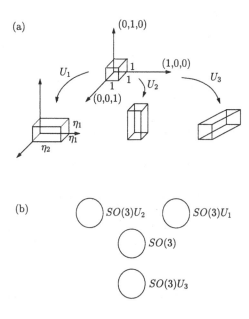

Figure 7: A cubic-to-tetragonal phase transformation. (a) Austenite and the three variants of martensite, (b) the corresponding energy wells $SO(3)$ and $SO(3)U_i$, $i = 1, 2, 3$.

for example, at $\theta = \theta_c$ there is one austenite energy well $SO(3)$ and three martensite energy wells $SO(3)U_i$ corresponding to the three possible *variants* of martensite. These energy wells are represented schematically by circles in Fig. 7.

By considering different rank-one connections between the different energy wells as explained in Section 3 we can construct different global energy minimisers which are combinations of variants or phases. To illustrate this let us suppose that $\theta = \theta_c$. Then it can be shown [14] that there is no rank-one connection between two matrices on a single well, nor between a martensite well and the austenite well (provided no $\eta_i = 1$, which we assume). However to each matrix on a martensite well there correspond two distinct matrices on each different martensite well. Once we have found such a rank-one connection between martensite wells

$$B - A = a \otimes n,$$

with $A \in SO(3)U_i, B \in SO(3)U_j, i \neq j$, we can construct a corresponding interface and even form a *simple laminate* consisting of the limit as $j \to \infty$ of a family of layers having normals n, alternating deformation gradients A, B and alternating thicknesses $\lambda/j, (1 - \lambda)/j$ (see Fig. 8).

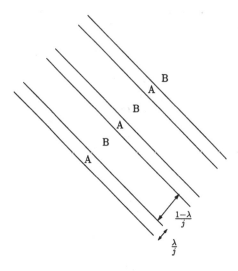

Figure 8: A simple laminate formed using the rank-one connection $B - A = a \otimes n$.

The Young measure corresponding to the sequence $z^{(j)} = Dy^{(j)}$ of deformation gradients generating this microstructure in the limit $j \to \infty$ is easily verified to be

$$\nu_x = \lambda \delta_A + (1 - \lambda)\delta_B,$$

where δ_A denotes the Dirac mass at A.

As another example we can consider a *double laminate* shown in Fig. 9 formed from four matrices A, B, C, D on martensite wells with rank-one connections

$$A - B = a \otimes n,$$
$$C - D = b \otimes m,$$
$$(B + \mu a \otimes n) - (D + \lambda b \otimes m) = c \otimes l,$$

and having the Young measure

$$\nu_x = \sigma(\mu \delta_A + (1 - \mu)\delta_B) + (1 - \sigma)(\lambda \delta_C + (1 - \lambda)\delta_D).$$

Note that since, for example, rank $(A - C) > 1$ in general, transition regions are required to interpolate between the single laminates, as in problem (5.1). The volume of these layers tends to zero as $j \to \infty$ and so the energy contribution due to the deformation gradients not belonging to $K(\theta_c)$ vanishes in the limit.

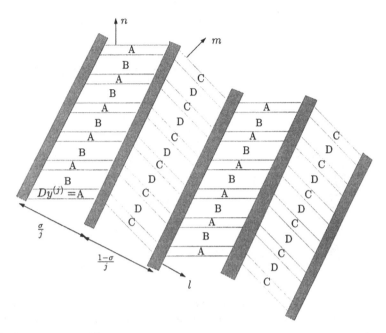

Figure 9: A double laminate formed from four matrices on martensite wells, showing the values of the deformation gradient $Dy^{(j)}$. The layers in which $Dy^{(j)}$ takes the values A, B, C, D have thicknesses $\frac{\mu}{j^2}$, $\frac{(1-\mu)}{j^2}$, $\frac{\lambda}{j^2}$, $\frac{(1-\lambda)}{j^2}$ respectively. The shaded areas are transition regions whose thicknesses are $O\left(\frac{1}{j^2}\right)$.

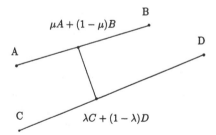

Figure 10: Diagram showing the rank-one connections in the double laminate shown in Fig. 9; each line represents a rank-one connection.

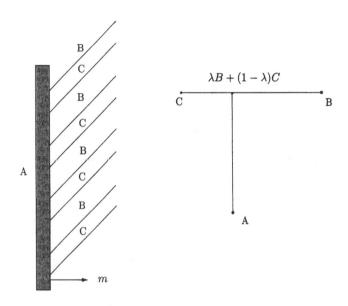

Figure 11: The austenite-martensite interface, and the corresponding rank-one connections. The matrix A is on the austenite well, B and C on different martensite wells. The corresponding Young measure is given by $\nu_x = \delta_A$ if $x \cdot m < \alpha$, and by $\nu_x = \lambda \delta_B + (1 - \lambda)\delta_C$ if $x \cdot m > \alpha$.

It is sometimes more convenient to show the different rank-one connections involved in a microstructure using a diagram such as Fig. 10 in which they are indicated by straight lines.

In general different sorts of microstructure involving laminates can be analysed by solving the algebraic problem of finding matrices on the energy-wells having the appropriate rank-one connections. Interesting examples are the *austenite-martensite interface*, in which a single laminate of martensite is connected along a planar interface to an austenite region (see Fig. 11 and [14]), and the *wedge* [22].

Clearly an important question for the analysis of microstructures is to classify the Young measures arising from sequences of gradients. Important necessary conditions are the *minors relations*. In general a minor $J = J(A)$ is a subdeterminant of an $m \times n$ matrix A. For example, in the case $m = n = 3$ there are 19 such minors in all, given by the elements of A, cof A and det A. If $J : M^{m \times n} \to \mathbf{R}$ is a minor and $(\nu_x)_{x \in \Omega}$ is the Young measure corresponding to a sequence $Dy^{(j)}$ of gradients satisfying suitable bounds (e.g. uniformly bounded in L^∞) then

$$J \left(\int_{M^{m \times n}} A \, d\nu_x(A) \right) = \int_{M^{m \times n}} J(A) \, d\nu_x(A) \qquad \text{a.e. } x \in \Omega.$$

In particular for $m = n = 3$

$$
\begin{aligned}
Dy(x) &= \int_{M^{3\times3}} A \, d\nu_x(A), \\
\operatorname{cof} Dy(x) &= \int_{M^{3\times3}} \operatorname{cof} A \, d\nu_x(A), \\
\det Dy(x) &= \int_{M^{3\times3}} \det A \, d\nu_x(A),
\end{aligned}
$$

where y denotes the weak limit of $y^{(j)}$. The minors relations are not sufficient for a parametrised measure $(\nu_x)_{x\in\Omega}$ to be a Young measure of gradients (c.f. [15, 23]). Necessary and sufficient conditions in terms of quasiconvexity have been obtained by Kinderlehrer & Pedregal [41], but on account of our lack of understanding of quasiconvexity these conditions are at present more of theoretical interest than a practical tool.

5.4 The two-well problem

It is impossible to adequately survey here the very active field of the analysis of crystal microstructure, and the reader should consult the cited references. In this section we consider briefly one special problem in which the issue of nonattainment of a minimum has been at least partially clarified. This is the *two-well problem* arising, for example, in orthorhombic to monoclinic transformations, in which $K = K(\theta)$ has the form

$$
K = SO(3)U_1 \cup SO(3)U_2
$$

for distinct positive definite symmetric matrices U_1, U_2. We suppose that the minimum value of $W(\cdot, \theta)$, attained on K, is zero. The case when $\det U_1 = \det U_2$ is discussed in [15], where it is shown that via a change of variables one can represent the above set K in the form

$$
K = SO(3)S^+ \cup SO(3)S^-,
$$

with $S^\pm = 1 \pm \delta e_3 \otimes e_1$, $\delta > 0$ and $\{e_1, e_2, e_3\}$ an orthonormal basis for $\mathbf{R}^3$. Consider the problem of minimising

$$
I(y) = \int_\Omega W(Dy, \theta) \, dx
$$

subject to the linear boundary conditions

$$
y|_{\partial\Omega} = Fx,
$$

where $F \in M_+^{3\times3}$. It is proved in [15] that the infimum of I is zero (i.e. there is a zero energy microstructure) if and only if $C = F^T F \in \mathcal{R}$, where $\mathcal{R}$ consists of symmetric matrices of the form

$$
\begin{bmatrix}
C_{11} & 0 & C_{13} \\
0 & 1 & 0 \\
C_{13} & 0 & C_{33}
\end{bmatrix}
$$

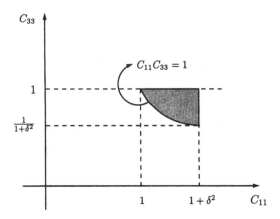

Figure 12: The set of strains C corresponding to zero-energy microstructures in the two-well problem; (C_{11}, C_{33}) must belong to the shaded region and $C_{13}^2 = C_{11}C_{33} - 1$.

with (C_{11}, C_{33}) in the region shown in Fig. 12 defined by the inequalities $C_{11} \le 1 + \delta^2$, $C_{33} \le 1$ and $C_{11}C_{33} \ge 1$, and with $C_{13}^2 = C_{11}C_{33} - 1$.

If $C \in \mathcal{R}$ satisfies either $C_{11} = 1 + \delta^2$, or $C_{33} = 1$ then F has the form

$$F = \lambda A + (1 - \lambda)B,$$

with $A \in SO(3)S^+$, $B \in SO(3)S^-$, $\lambda \in [0, 1]$, and the Young measure corresponding to the sequence $Dy^{(j)}$ for *any* minimising sequence $y^{(j)}$ is given by

$$\nu_x = \lambda \delta_A + (1 - \lambda)\delta_B.$$

In particular, if $\lambda \ne 0, 1$ (i.e. if $F \notin K$) then the Young measure is not a Dirac mass and so the *minimum is not attained*. Recent work of Müller and Šverák [50] suggests that, surprisingly, the minimum may be attained if C takes other values in $\mathcal{R}$, though by a very irregular deformation.

Acknowledgements

I thank Ali Taheri for preparing a draft of the notes of these lectures and Tanya Smekal for providing the figures. I am grateful to EU (ERBSCI** CT000670) and EPSRC (GR/J03466) for financial support.

References

[1] E Acerbi and N Fusco. Semicontinuity problems in the calculus of variations. *Arch. Rat. Mech. Anal.* **86**, 125-145, 1984.

[2] R Adams. *Sobolev Spaces.* Academic Press, 1975.

[3] L Ambrosio. Variational problems in SBV. *Acta Applicandae Mathematicae* **17**, 1-40, 1989.

[4] L Ambrosio. Existence theory for a new class of variational problems. *Arch. Rat. Mech. Anal.* **111**, 291-322, 1990.

[5] L Ambrosio and D Pallara. Partial regularity of free discontinuity sets I. Preprint.

[6] L Ambrosio, N Fusco, and D Pallara. Partial regularity of free discontinuity sets II. Preprint.

[7] E J Balder. Lectures on Young measures. *Cahiers de Mathématiques de la Décision of CEREMADE*, to appear.

[8] J M Ball. Convexity conditions and existence theorems in nonlinear elasticity. *Arch. Rat. Mech. Anal.* **63**, 337-403, 1977.

[9] J M Ball. Strict convexity, strong ellipticity, and regularity in the calculus of variations. *Proc. Camb. Phil. Soc.* **87**, 501-513, 1980.

[10] J M Ball. Global invertibility of Sobolev functions and the interpenetration of matter. *Proc. Roy. Soc. Ed. A* **88**, 315-328, 1981.

[11] J M Ball. Discontinuous equilibrium solutions and cavitation in nonlinear elasticity. *Phil. Trans. Roy. Soc. Lond. A* **306**, 557-611, 1982.

[12] J M Ball. Minimisers and the Euler-Lagrange equations. In *Proc. ISIMM conference, Paris.* Springer-Verlag, 1983.

[13] J M Ball. A version of the fundamental theorem for Young measures. In M Rascle, D Serre, and M Slemrod, editors, *Proceedings of conference on 'Partial Differential Equations and Continuum Models of Phase Transitions'*, Springer Lecture Notes in Physics No. 359, 3-16, 1989.

[14] J M Ball and R D James. Fine phase mixtures as minimisers of energy. *Arch. Rat. Mech. Anal.* **100**, 13-52, 1987.

[15] J M Ball and R D James. Proposed experimental tests of a theory of fine microstructure, and the two-well problem. *Phil. Trans. Roy. Soc. Lond. A* **338**, 389-450, 1992.

[16] J M Ball and V J Mizel. One-dimensional variational problems whose minimisers do not satisfy the Euler-Lagrange equations. *Arch. Rat. Mech. Anal.* **90**, 325-388, 1985.

[17] J M Ball and F Murat. $W^{1,p}$-quasiconvexity and variational problems for multiple integrals. *J. Funct. Anal.* **58**, 225-253, 1984.

[18] P Bauman, N C Owen and D Phillips. Maximal smoothness of solutions to certain Euler-Lagrange equations from nonlinear elasticity. *Proc. Roy. Soc. Ed. A* **119**, 241-263, 1991.

[19] P Bauman, N C Owen and D Phillips. Maximum principles and a priori estimates for a class of problems from nonlinear elasticity. *Annales de l'Institut Henri Poincaré - Analyse non linéaire*, **8**, 119-157, 1991.

[20] P Bauman, N C Owen and D Phillips. Maximum principles and a priori estimates for an incompressible material in nonlinear elasticity. *Comm. in Partial Diff. Eqns* **17**, 1185-1212, 1992.

[21] P Bauman and D Phillips. Univalent minimisers of polyconvex functionals in 2 dimensions. *Arch. Rat. Mech. Anal.* **126**, 161-181, 1994.

[22] K Bhattacharya. Wedge-like microstructure in martensites. *Acta Metallurgica et Materialia* **39**, 2431-2444, 1991.

[23] K Bhattacharya, N B Firoozye, R D James and R V Kohn. Restrictions on microstructure. *Proc. Roy. Soc. Ed. A* **124**, 843-878, 1994.

[24] A Braides. Homogenisation of bulk and surface energies. *Arch. Rat. Mech. Anal.*, to appear.

[25] H Brezis. *Analyse Fonctionelle*. Masson, 1987.

[26] P G Ciarlet and J Nečas. Unilateral problems in nonlinear three-dimensional elasticity. *Arch. Rat. Mech. Anal.* **87**, 319-338, 1985.

[27] B Dacorogna and J-P Haeberly. Remarks on a numerical study of convexity, quasiconvexity, and rank one convexity. Preprint.

[28] B Dacorogna and J-P Haeberly. Some numerical methods for the study of the convexity notions arising in the calculus of variations. Preprint.

[29] G David and S Semmes. On the singular set of minimisers of the Mumford-Shah functional. To appear.

[30] E De Giorgi and L Ambrosio. Un nuovo tipo di funzionale del calcolo delle variazioni. *Atti Accad. Naz. Lincei Cl. Sci. Fis. Mat. Nat.* (8)**82**, 199-210, 1988.

[31] L C Evans. Quasiconvexity and partial regularity in the calculus of variations. *Arch. Rat. Mech. Anal.* **95**, 227-268, 1986.

[32] L C Evans and R Gariepy. *Measure Theory and Fine Properties of Functions.* CRC Press, 1992.

[33] I Fonseca and W Gangbo. Local invertibility of Sobolev functions. *SIAM J. Math. Anal.* **26**, 280-304, 1995.

[34] N Fusco and J Hutchinson. Partial regularity in problems motivated by nonlinear elasticity. *SIAM J. Math. Anal.* **22**, 1516-1551, 1991.

[35] N Fusco and J Hutchinson. Partial regularity and everywhere continuity for a model problem from nonlinear elasticity. *J. Australian Math. Soc. A* **57**, 158-169, 1994.

[36] A N Gent and P B Lindley. Internal rupture of bonded rubber cylinders in tension. *Proc. Roy. Soc. Lond. A* **249**, 195-205, 1958.

[37] A N Gent and B Park. Failure processes in elastomers at or near a rigid spherical inclusion. *J. Mat. Sci.* **19**, 1947-1956, 1984.

[38] M Giaquinta, G Modica, and J Souček. Cartesian currents, weak diffeomorphisms and existence theorems in nonlinear elasticity. *Arch. Rat. Mech. Anal.* **106**, 97-159, 1989. Addendum, *ibid.* **109**, 385-392, 1990.

[39] C O Horgan and D A Polignone. Cavitation in nonlinearly elastic solids: a review. *Appl. Mech. Rev.* **48**, 471-485, 1995.

[40] R D James and S J Spector. The formation of filamentary voids in solids. *J. Mech. Phys. Solids* **39**, 783-813, 1991.

[41] D Kinderlehrer and P Pedregal. Characterisations of Young measures generated by gradients. *Arch. Rat. Mech. Anal.* **115**, 329-365, 1991.

[42] J Kristensen. Thesis, Technical University of Lyngby, 1994.

[43] P Marcellini. Approximation of quasiconvex functions, and lower semicontinuity of multiple integrals. *Manuscripta Math.* **85**, 1-28, 1985.

[44] G H Meisters and C Olech. Locally one-to-one mappings and a classical theorem on Schlicht functions. *Duke Math. J.* **30**, 63-80, 1963.

[45] C B Morrey. Quasi-convexity and the lower semicontinuity of multiple integrals. *Pacific J. Math.* **2**, 25-53, 1952.

[46] C B Morrey. *Multiple Integrals in the Calculus of Variations.* Springer, 1966.

[47] S Müller. Weak continuity of determinants and nonlinear elasticity. *Comptes Rendus Acad. Sciences Paris* **307**, 501-506, 1988.

[48] S Müller, T. Qi and B S Yan. On a new class of elastic deformations not allowing for cavitation. *Ann. Inst. Henri Poincaré,Analyse Nonlinéaire* **11**, 217-243, 1994.

[49] S Müller and S J Spector. An existence theory for nonlinear elasticity that allows for cavitation. *Arch. Rat. Mech. Anal.* to appear.

[50] S Müller and V Šverák. Surprising attainment results for the two-well problem. Preprint SFB256, Bonn University.

[51] D Mumford and J Shah. Optimal approximation by piecewise smooth functions and associated variational problems. *Comm. Pure Appl. Math.* **17**, 577-685, 1989.

[52] J Nečas. *Les Méthodes Directes en Théorie des Équations Elliptiques.* Academia, Prague, 1967.

[53] J Nečas. Example of an irregular solution to a nonlinear elliptic system with analytic coefficients and conditions for regularity. In *Theory of Nonlinear Operators*, Akademie-Verlag, Berlin, 197-206, 1977.

[54] J Nečas. *Introduction to the Theory of Nonlinear Elliptic Equations.* Tuebner, Leipzig, 1983.

[55] A E Oberth and R S Bruenner. Tear phenomena around solid inclusions in castable elastomers. *Trans. Soc. Rheol.* **9**, 165-185, 1965.

[56] R W Ogden. Large deformation isotropic elasticity - on the correlation of theory and experiment for incompressible rubberlike solids. *Proc. Roy. Soc. Lond. A* **326**, 562-584, 1972.

[57] R W Ogden. Large deformation isotropic elasticity: on the correlation of theory and experiment for compressible rubberlike solids. *Proc. Roy. Soc. Lond. A* **328**, 567-583, 1972.

[58] K Ohtsuka. Mathematical aspects of fracture mechanics. *Lecture Notes in Num. Appl. Anal.* **13**, 39-59, 1994.

[59] G P Parry. On the planar rank-one convexity condition. *Proc. Roy. Soc. Ed. A* **125**, 247-264, 1995.

[60] P Pedregal. Some remarks on quasiconvexity and rank-one convexity. *Proc. Roy. Soc. Ed.* to appear.

[61] J Sivaloganathan. Uniqueness of regular and singular equilibria for spherically symmetric problems of nonlinear elasticity. *Arch. Rat. Mech. Anal.* **96**, 97-136, 1986.

[62] R Stringfellow and R Abeyaratne. Cavitation in an elastomer - comparison of theory with experiment. *Materials Science and Engineering A - Structural Materials Properties, Microstructure and Processing* **112**, 127-131, 1989.

[63] C A Stuart. Radially symmetric cavitation for hyperelastic materials. *Ann. Inst. H. Poincaré. Anal. Non. Linéaire*, **2**, 33-66, 1985.

[64] V Šverák. Lower-semicontinuity of variational integrals and compensated compactness. In *Proc. International Congress of Mathematicians, Zurich 1994*. To appear.

[65] V Šverák. Regularity properties of deformations with finite energy. *Arch. Rat. Mech. Anal.* **100**, 105-127, 1988.

[66] V Šverák. Rank-one convexity does not imply quasiconvexity. *Proc. Roy. Soc. Ed. A* **120**, 185-189, 1992.

[67] M Valadier. A course on Young measures. *Workshop on Measure Theory and Real Analysis, Grado 1993*, 1994. Preprint.

[68] A Weinstein. A global invertibility theorem for manifolds with boundary. *Proc. Roy. Soc. Ed. A* **99**, 283-284, 1985.

[69] L C Young. *Lectures on the Calculus of Variations and Optimal Control Theory*. Saunders, 1969. Reprinted by Chelsea 1980.

Analysis of the $2d$ and $3d$ Navier-Stokes Equations

J.D. Gibbon

Department of Mathematics,
Imperial College of Science, Technology and Medicine,
London

1   Introduction

One of the most puzzling problems in theoretical fluid mechanics is the behaviour of the $3d$ Navier-Stokes equations, not least because they hold the key to a proper understanding of fully developed turbulence. Analysts have been unable to prove that solutions of the $3d$ Navier-Stokes equations remain regular for all finite times [7, 10, 20, 25, 26] while in tandem, computational fluid dynamicists have been unable, in a consistent fashion, to resolve these solutions for arbitrarily large Reynolds numbers. Both of these difficulties point to some inadequacy in our understanding of the equations themselves. The $2d$ Navier-Stokes equations are well understood, but the existence of finite time singularities in solutions of the $3d$ Navier-Stokes equations is still an open question because the occurrence of these cannot be discounted. Nevertheless, it is generally believed by most (although not all) in these fields that these equations are regular for all finite times, but the mechanism by which they may remain so has yet to be discovered. (In contrast, it is generally believed that the $3d$ Euler equations do exhibit finite time singularities. Two reviews of these issues for the $3d$ Euler equations can be found in [22, 23].) Despite this lack of a complete understanding, there is much that can be understood. Moreover, we can examine what is not understood to point us to where the problems lie.

One of the ways that modern dynamical systems theory has been applied to infinite dimensional problems is in the development and application of the idea of the global attractor $\mathcal{A}$ for dissipative systems described by parabolic partial differential equations [27]. In particular, the sharp estimate of the Lyapunov dimension d_{att} of the global attractor $\mathcal{A}$ of the $2d$ Navier-Stokes equations found by Constantin, Foias & Temam [7, 8, 27] can be thought of as one of the great successes in this area. Their result in [8] was that $d_{att}(\mathcal{A})$ is bounded above by

$$d_{att} \leq c\,\mathcal{G}^{2/3}\left(1 + \log \mathcal{G}\right)^{1/3}, \qquad (1.1)$$

where $\mathcal{G}$ is the $2d$ Grashof number which will be defined later. If one thinks of $\mathcal{N}$ as the number of degrees of freedom of the system, L as the system length, d as the number of spatial dimensions and ℓ as a natural small length scale in the dynamics from all smooth initial conditions, then Landau loosely related these by

$$d_{att} \sim \left(\frac{L}{\ell}\right)^d. \tag{1.2}$$

If we identify d_{att} with $\mathcal{N}$ then this connects the attractor dimension with small length scales. $\mathcal{N}$ could also be thought of as the number of grid points that would be needed in a numerical integration to resolve the smallest features in the flow. One of the merits of this definition is that it uses not just the size of the dynamical system, but also the effect of the dynamics.

It is often the case that Navier-Stokes analysts use periodic boundary conditions. Of course, real world fluid problems generally have boundaries but in the regularity problem these are not the main issue: there are fundamental problems with the periodic problem before one even considers boundaries. For this reason it is easier to take the problem stripped down to its minimum and consider the periodic case. Let us therefore take the forced d-dimensional incompressible Navier-Stokes equations ($d = 2, 3$) with periodic boundary conditions on the domain $\Omega \equiv [0, L]^d$

$$\frac{\partial \mathbf{u}}{\partial t} + \mathbf{u} \cdot \nabla \mathbf{u} = \nu \Delta \mathbf{u} - \nabla p + \mathbf{f}, \tag{1.3}$$

where ν is the viscosity, $\mathbf{f}(\mathbf{x})$ is the applied (divergence free) body force and the pressure p is determined by the condition of incompressibility $\nabla \cdot \mathbf{u} = 0$. Without loss of generality we also take the flow to be mean zero so that $\int_\Omega \mathbf{u} \, d^d x = 0$. The evolution equation for the vorticity $\boldsymbol{\omega} = \text{curl}\,\mathbf{u}$ is

$$\frac{\partial \boldsymbol{\omega}}{\partial t} + \mathbf{u} \cdot \nabla \boldsymbol{\omega} - \boldsymbol{\omega} \cdot \nabla \mathbf{u} = \nu \Delta \boldsymbol{\omega} + \text{curl}\,\mathbf{f}. \tag{1.4}$$

Alternatively, the pressure obeys a Poisson equation which can be obtained by taking the divergence of (1.3), ignoring the forcing, and using the divergence free condition on $\mathbf{u}$, giving

$$\Delta p = -\sum_{i,j} u_{i,j}\, u_{j,i}. \tag{1.5}$$

Solving this equation for p at each step is a major task and it is here where many of the computing problems lie. Let us therefore consider the energy H_0 and the enstrophy H_1 (see also Appendix A), defined by

$$H_0 = \int_\Omega |\mathbf{u}|^2 \, d^d x, \qquad H_1 = \frac{d}{dt} \int_\Omega |\boldsymbol{\omega}|^2 \, d^d x, \tag{1.6}$$

and see how they evolve. H_0 satisfies

$$\frac{1}{2}\dot{H}_0 = -\nu \int_\Omega |\nabla \mathbf{u}|^2 \, d^d x - \int_\Omega \mathbf{u} \cdot (\mathbf{u} \cdot \nabla)\mathbf{u} \, d^d x + \int_\Omega \mathbf{u} \cdot \mathbf{f} \, d^d x - \int_\Omega \mathbf{u} \cdot \nabla p \, d^d x. \tag{1.7}$$

The second term on the right hand side can be written as $\mathbf{u} \cdot (\mathbf{u} \cdot \nabla)\mathbf{u} = \text{div}(\frac{1}{2}u^2\mathbf{u}) - \frac{1}{2}u^2\text{div}\,\mathbf{u}$. The last term in this expression is zero and the first vanishes by the Divergence Theorem. Moreover, an integration by parts in the pressure term shows that this term is also zero as $\int_\Omega u_{i,i}p\,d^dx = 0$. It also shows why this term does *not* vanish if one considers $\mathbf{u}$ in a space higher than L^2. Having $\mathbf{u}$ bounded only in L^2 but no higher L^p norm is manifestly a serious restriction which has dire consequences in the $3d$ problem. We are finally left with

$$\frac{1}{2}\dot{H}_0 \leq -\nu H_1 + H_0^{1/2}\|\mathbf{f}\|_2. \tag{1.8}$$

Because we are dealing with mean zero periodic functions, Poincaré's inequality is applicable [10] and so (1.8) becomes

$$\frac{1}{2}\dot{H}_0 \leq -\nu k_0^2 H_0 + H_0^{1/2}\|\mathbf{f}\|_2, \tag{1.9}$$

where $k_0 = 2\pi/L$ is the smallest nonzero wavenumber on Ω. Hence, for arbitrarily large initial values, H_0 is bounded for all t as

$$\varlimsup_{t\to\infty} H_0 \leq \frac{\|\mathbf{f}\|_2^2}{\nu^2 k_0^4} = \frac{\nu^2}{L^{6-d}k_0^4}\mathcal{G}^2, \tag{1.10}$$

where the d-dimensional $(d = 2,3)$ Grashof number is defined by

$$\mathcal{G} = L^{3-d/2}\|\mathbf{f}\|_2/\nu^2. \tag{1.11}$$

Moreover, using the time averaging operation

$$\langle g \rangle = \varlimsup_{\substack{t\to\infty \\ g(0)}} \sup \frac{1}{t}\int_0^t g(\tau)\,d\tau, \tag{1.12}$$

(1.10) shows that (see [21])

$$\langle H_1 \rangle \leq \frac{\|\mathbf{f}\|_2^2}{\nu^2 k_0^2} = \frac{\nu^2}{L^{6-d}k_0^2}\mathcal{G}^2, \tag{1.13}$$

which is one way of expressing Leray's inequality. The salient point here is that having $\mathbf{u}$ bounded in L^2 in (1.10) gives us little control over $\mathbf{u}$ which may still exhibit singularities. Furthermore, while the time average of H_1 is controlled from above this does *not* imply that H_1 itself exhibits no time singularities. It is for these reasons that Leray introduced his work on weak solutions in the 1930's [7, 10, 21, 26]. Nevertheless, since H_0 is the energy in the system, it allows us to compute the energy dissipation rate $\varepsilon = \nu\langle H_1\rangle L^{-3}$ which is bounded for the whole motion. From this we can compute a "length", known as the Kolmogorov length λ_K [18], which comes from dimensional arguments as a combination of ε and ν

$$\lambda_K^{-1} = \left(\frac{\varepsilon}{\nu^3}\right)^{1/4}. \tag{1.14}$$

This is a result about time averages, not about pointwise behaviour in time. To find the latter we must look at the vorticity version of (1.3), where we find that the enstrophy H_1 obeys

$$\frac{1}{2}\dot{H}_1 = -\nu \int_\Omega |\nabla\boldsymbol{\omega}|^2 \, d^d x + \int_\Omega \boldsymbol{\omega}\cdot(\boldsymbol{\omega}\cdot\nabla)\,\mathbf{u}\, d^d x + \int_\Omega \boldsymbol{\omega}\cdot\operatorname{curl}\mathbf{f}\, d^d x, \quad (1.15)$$

where we have integrated away one term as in (1.7) and (1.8). Now, in $3d$, the term $\int_\Omega \boldsymbol{\omega}\cdot\boldsymbol{\omega}\cdot\nabla\mathbf{u}\, d^d x$ in (1.15) is called the vortex stretching term. It is this that does the damage, creating or destroying vorticity and stretching it so that the vorticity gradients can become very large. In $2d$ this term is zero as the vorticity vector is perpendicular to the plane. It is in this term that the difference between $2d$ and $3d$ flows lies and this difference is fundamental and far reaching. For the $2d$ case, (1.15) can be rewritten by integrating the forcing term by parts and splitting the two terms up using Hölder's inequality to give

$$\frac{1}{2}\dot{H}_1 = -\frac{\nu}{2}\int_\Omega |\nabla\boldsymbol{\omega}|^2 \, d^d x + \frac{1}{2\nu}\|\mathbf{f}\|_2^2. \quad (1.16)$$

The vortex stretching term is now missing and a time averaging allows us to define the bounded quantity known as the enstrophy dissipation rate χ by

$$\chi = \nu L^{-2}\left\langle \|\nabla\omega\|_2^2 \right\rangle, \quad (1.17)$$

which, as we can see from (1.15), is a bounded quantity. In parallel with the Kolmogorov length λ_K which was formed out of ε and ν, this defines the Kraichnan length λ_{K_r} [19], which is defined as

$$\lambda_{K_r}^{-1} = \left(\frac{\chi}{\nu^3}\right)^{1/6}. \quad (1.18)$$

One may also notice that an integration by parts produces

$$\int_\Omega |\boldsymbol{\omega}|^2 \, d^2 x \leq \left(\int_\Omega |\nabla\boldsymbol{\omega}|^2 \, d^2 x\right)\left(\int_\Omega |\mathbf{u}|\, d^2 x\right), \quad (1.19)$$

thereby enabling us to rewrite (1.16) as

$$\dot{H}_1 \leq -\nu\frac{H_1^2}{H_0} + \nu^{-1}\|\mathbf{f}\|_2^2. \quad (1.20)$$

Because H_0 is bounded above for all t, H_1 therefore contracts into an absorbing ball for arbitrarily large initial values

$$\varlimsup_{t\to\infty} H_1 \leq \frac{\nu^2}{L^4 k_0^2}\mathcal{G}^2. \quad (1.21)$$

In addition to H_0, finding an upper bound for H_1 for all t is a major, in fact *the* major step, in proving regularity. That it works for the $2d$ case is dependent on the absence of the vortex stretching term. If one includes this term for the $3d$ case then we get

$$\dot{H}_1 \leq -\nu \frac{H_1^2}{H_0} + c\|\boldsymbol{\omega}\|_\infty H_1 + \nu^{-1}\|\mathbf{f}\|_2^2. \tag{1.22}$$

The main point now is to see how we may deal with the full $3d$ system with the $\|\boldsymbol{\omega}\|_\infty$ term. The next section is devoted to something called a "Ladder Theorem" which is valid for both dimensions.

2 The Navier-Stokes Ladder Theorem

In Section 1, the energy $H_0 = \int_\Omega |\mathbf{u}|^2 \, d^d x$ was shown to be bounded from above *a priori* in every dimension. Unfortunately, this tells us little about the velocity field $\mathbf{u}(\mathbf{x}, t)$ because functions bounded in L^2 can still display spatial singularities. To obtain more information about how the derivatives of the velocity and/or vorticity fields might be controlled, we need to consider the seminorms H_n defined by

$$H_n = \int_\Omega |D^n \mathbf{u}|^2 \, d^d x, \tag{2.1}$$

and how they behave for arbitrary t. D^n in (2.1) is a tensor quantity which, in effect, means every derivative of order n while the case $n = 1$ coincides with the gradient operator. In fact, it is not difficult to show that for divergence-free vector fields on a periodic domain, $\int_\Omega |D^n \mathbf{u}|^2 \, d^d x = \int_\Omega |\text{curl}^n \mathbf{u}|^2 \, d^d x$. As equation (1.13) has shown, the only general information we have about the weak solutions is that the time integral of H_1, is bounded from above *a priori*. This again does not tell us much about H_1 itself as temporal singularities in this seminorm could still occur.

It is clearly desirable to have a result which relates H_n to H_{n-s} (say) for general n with $1 \leq s \leq n$. To this end, our aim in this section is to display a theorem of this type. It is also necessary not only to show how the H_n evolve in time but also how the body forcing affects the flow. This theorem we will refer to as the "Navier-Stokes Ladder Theorem" or just the "Ladder Theorem" and its proof can be found in [2, 3, 10]. The main tools which are used in the proof of this theorem are the Divergence Theorem, Cauchy's and Schwarz's inequalities, and finally the calculus inequalities of Gagliardo and Nirenberg [1, 10, 24]. Let the set of seminorms associated with the forcing be defined by

$$\Phi_n = \int_\Omega |D^n \mathbf{f}|^2 \, d^d x, \tag{2.2}$$

where $\nabla \cdot \mathbf{f} = 0$ without loss of generality. Before we state this theorem it is necessary to make some comments on how the flow is forced. We assume a forcing which is independent of time, and in which the forcing function $\mathbf{f}(\mathbf{x})$

is assumed to have a cutoff in its spectrum such that it has a smallest length scale

$$\lambda_f^{-2} = \sup_n \left(\frac{\Phi_{n+1}}{\Phi_n} \right). \tag{2.3}$$

In other words, the cut-off k_f in the spectrum is given by $k_f = 2\pi/\lambda_f$. It is also convenient to introduce a natural 'time' $\tau = L^2\nu^{-1}$ for the system based upon the length of the box L and the viscosity ν. The forcing function $\mathbf{f}$ is dimensionally an acceleration so $\mathbf{u}_f = \tau\mathbf{f}$ is dimensionally a velocity vector. Now we can define

$$F_n = H_n + \tau^2\Phi_n = H_n + \int_\Omega |D^n\mathbf{u}_f|^2 \, d^d x. \tag{2.4}$$

The F_n not only contains n derivatives on the 3 components of velocity but also the 3 components of the forcing. We define an inverse squared length scale as a combination of the inverse squared box length and inverse squared forcing length by

$$\lambda_0^{-2} = \lambda_f^{-2} + L^{-2}. \tag{2.5}$$

We now have the following result.

Theorem 1 *In $d = 2$ and $d = 3$ dimensions and for $1 \leq s \leq n$,*

$$\frac{1}{2}\dot{F}_n \leq -\nu\frac{F_n^{1+1/s}}{F_{n-s}^{1/s}} + \left(c_n \|D\mathbf{u}\|_\infty + \nu\lambda_0^{-2}\right) F_n. \tag{2.6}$$

This is the Ladder Theorem, the proof of which is given in [2, 3, 10]. In the penultimate step of the proof of Theorem 1, the result stands as

$$\frac{1}{2}\dot{F}_n \leq -\nu F_{n+1} + \left(c_n \|D\mathbf{u}\|_\infty + \nu\lambda_0^{-2}\right) F_n, \tag{2.7}$$

which has an equivalent lower bound of

$$\frac{1}{2}\dot{F}_n \geq -\nu F_{n+1} - \left(c_n \|D\mathbf{u}\|_\infty + \nu\lambda_0^{-2}\right) F_n. \tag{2.8}$$

The first term on the right hand side of (2.6) comes from (2.7) by using the inequality $F_n^{s+1} \leq F_{n+1}^s F_{n-s}$. The objects

$$\kappa_{n,r}(t) = \left(\frac{F_n}{F_r} \right)^{\frac{1}{2(n-r)}} \tag{2.9}$$

can be thought of as the natural time dependent wavenumbers for the system [3]. They are ordered such that $\kappa_{n,r} \leq \kappa_{m,r}$ for $n < m$ and the smallest $\kappa_{1,0}$ is bounded from below. Simply by dividing through (2.6) by F_n and using the time averaging operation defined in (1.12) gives the result for $n \geq 2$

$$\left\langle \kappa_{n,r}^2 \right\rangle \leq c_{n,r} \nu^{-1} \left\langle \|D\mathbf{u}\|_\infty \right\rangle + \lambda_0^{-2}, \tag{2.10}$$

which can be found in [3]. (Note that $\left\langle \kappa_{1,0}^2 \right\rangle$ is controlled but this is just another way of expressing Leray's inequality.) How far one can then go in estimating the right hand side of (2.10) is related to the regularity problem for the Navier-Stokes equations. Moreover, the $\kappa_{n,r}$ are also subject to a 'ladder' theorem, as follows.

Theorem 2 *For* $s = n - r$ *the* $\kappa_{n,r}$ *satisfy*

$$s\dot{\kappa}_{n,r} \leq -\nu\kappa_{n+1,r}^3 + \nu\kappa_{r+1,r}^2 \kappa_{n,r} + c_{n,r} \left(\|D\mathbf{u}\|_\infty + \nu\lambda_0^{-2} \right) \kappa_{n,r}. \tag{2.11}$$

Proof: This proceeds by changing the inequality for F_n in (2.6) to one for $\kappa_{n,r}$. Let $s = n - r$, then

$$2\kappa_{n,r}^{2s-1} \dot{\kappa}_{n,r} = \frac{\dot{F}_n}{F_r} - \frac{\dot{F}_r}{F_r} \kappa_{n,r}^{2s} \leq \frac{\dot{F}_n}{F_r} - \frac{\dot{F}_r}{F_r} \kappa_{n,r}^{2s}. \tag{2.12}$$

Using (2.7) for $\dot{F}_n$ and then again with n replaced by r but using the lower bound (2.8), (2.12) simplifies to

$$s\kappa_{n,r}^{2s-1} \dot{\kappa}_{n,r} \leq -\nu\kappa_{n+1,r}^{2(s+1)} + \left(c_{n,r} \|D\mathbf{u}\|_\infty + \nu\lambda_0^{-2} \right) \kappa_{n,r}^{2s} + \nu\kappa_{r+1,r}^2 \kappa_{n,r}^{2s}, \tag{2.13}$$

which, because $\kappa_{n,r} \leq \kappa_{n+1,r}$, gives the result.

Estimates for $\langle \kappa_{n,r} \rangle$ and $\left\langle \kappa_{n+1,r}^3 \right\rangle$ are important as they hold the key to several other quantities. Basic known Navier-Stokes results which we require, such as upper bounds on $\langle F_1 \rangle$ and $\langle F_2 \rangle$, can be found in Appendix A along with the definitions of the $2d$ and $3d$ Grashof numbers $\mathcal{G}$ which are used as the basic parameters of the problem. It is also possible to find a ladder and estimate length scales for the Boussinesq equations of convection [15]. The results obtained so far are summarised in Table 1.

Definition of F_n	$F_n = \left(\|D^n\mathbf{u}\|_2^2 + \|D^N\mathbf{u}_f\|_2^2 \right)$
Ladder for F_n	$\frac{1}{2}\dot{F}_n \leq -\nu\frac{F_n^{1+1/s}}{F_{n-s}^{1/s}} + \left(c\|D\mathbf{u}\|_\infty + \nu\lambda_0^{-2} \right) F_n$
Definition of $\kappa_{n,r}$	$\kappa_{n,r} = \left(\frac{F_n}{F_r} \right)^{\frac{1}{2(n-r)}}, \quad r < n$
Time average of $\kappa_{n,r}^2$	$\left\langle \kappa_{n,r}^2 \right\rangle \leq c_{n,r} \nu^{-1}\langle \|D\mathbf{u}\|_\infty \rangle + \lambda_0^{-2}$
Ladder for $\kappa_{n,r}$	$s\dot{\kappa}_{n,r} \leq -\nu\kappa_{n+1,r}^3 + \nu\kappa_{r+1,r}^2 \kappa_{n,r}$ $+ \left(c\|D\mathbf{u}\|_\infty + \nu\lambda_0^{-2} \right) \kappa_{n,r}$

Table 1: A summary of results.

3   The 2d Navier-Stokes Equations

The Ladder Theorem expressed in Theorem 1, and summarised in Table 1, can now be used to see if it is possible to prove that there is an absorbing ball for *each* of the F_n. The main piece of extra information that we can use in the 2d case is that we have an absorbing ball for F_1. The various results found so far, expressed in terms of the F_n instead of the H_n, can be found in Table 2 and are derived in Appendix A.

The main task here is to gain control over the $\|Du\|_\infty$ term. For each component of **u** we use a 2d calculus inequality

$$\|Du_i\|_\infty \le c \, \|D^n u_i\|_2^a \, \|Du_i\|_2^{1-a}, \tag{3.1}$$

where $a = (n-1)^{-1}$ and $n \ge 3$. In turn this gives

$$\|D\mathbf{u}\|_\infty \le c \, F_n^{a/2} \, F_1^{(1-a)/2}. \tag{3.2}$$

Choosing $s = n - 1$ in the ladder, we obtain

$$\frac{1}{2}\dot{F}_n \le -\nu \, \frac{F_n^{1+\frac{1}{n-1}}}{F_1^{\frac{1}{n-1}}} + \left(c_n \, F_n^{a/2} F_1^{(1-a)/2} + \nu \lambda_0^{-2} \right) F_n. \tag{3.3}$$

It is clear from (3.3) that the exponent of the negative term is stronger than the that of the nonlinear term enabling us to find an absorbing ball. The result of this is

$$\varlimsup_{t \to \infty} F_n \le c_n \, \nu^{-2(n-1)} \left(\varlimsup_{t \to \infty} F_1 \right)^n + \lambda_0^{-2(n-1)} \left(\varlimsup_{t \to \infty} F_1 \right). \tag{3.4}$$

2d Grashof number	$\mathcal{G} = L^2 \nu^{-2} \|\mathbf{f}\|_2$
Absorbing ball for F_0	$\varlimsup_{t\to\infty} F_0 \le c \, \nu^2 \mathcal{G}^2$
Absorbing ball for F_1	$\varlimsup_{t\to\infty} F_1 \le c \, \lambda_0^{-2} \nu^2 \mathcal{G}^2$
Time average of F_1	$\langle F_1 \rangle \le \lambda_0^{-2} \nu^2 \mathcal{G}^2$
Time average of F_2	$\langle F_2 \rangle \le \lambda_0^{-4} \nu^2 \mathcal{G}^2$
Time average of $\kappa_{1,0}^2$	$\langle \kappa_{1,0}^2 \rangle \le \lambda_0^{-2}$
Time average of $\kappa_{2,1}^2$	$\langle \kappa_{2,1}^2 \rangle \le \lambda_0^{-2}$
Absorbing ball for F_n	$\varlimsup_{t\to\infty} F_n \le c_4 \, \lambda_0^{-2n} \nu^2 \left(\mathcal{G}^{2n} + \mathcal{G}^2 \right)$

Table 2: Results for the 2d case.

Now from Table 2, we have an estimate for $\overline{\lim}_{t\to\infty} F_1$ in terms of the $2d$ Grashof number $\mathcal{G}$ which, when substituted into (3.4), gives

$$\overline{\lim_{t\to\infty}} F_n \le c\,\lambda_0^{-2n}\nu^2\left(\mathcal{G}^{2n} + \mathcal{G}^2\right). \tag{3.5}$$

Obviously, the bound on F_1 implies the bound on F_n given in (3.5), so no singularities in any derivative can develop from smooth initial data in a $2d$ flow. This, then, is a nice result which is gratifying as an exercise in understanding the $2d$ Navier-Stokes equations as a set of PDE's. The triumph is only minor, however, as $2d$ flows are not particularly physical and their inability to create or destroy vorticity due to the absence of the vortex stretching term makes them a poor substitute for $3d$ flows. As we shall see in later sections, it is the control over F_1 around which the difference between the $2d$ and $3d$ cases revolves.

Because the $\kappa_{n,r}$ operate like a spectrum of wavenumbers, the $\kappa_{n,r}$ have more physical meaning than the F_n. It is from inequality (2.11) that most of the $2d$ and $3d$ Navier-Stokes results can be derived. Now we introduce the quantities

$$\mathcal{N}_{n,r} = \lambda_0^2 \kappa_{n,r}^2. \tag{3.6}$$

The time averages $\langle \mathcal{N}_{n,r} \rangle = \lambda_0^2 \langle \kappa_{n,r}^2 \rangle \sim (\lambda_0/\ell)^2$ play the role of $\mathcal{N}$ in (1.2) for a $2d$ system with λ_0 replacing L. In $3d$ our interest lies in $\langle \kappa_{n,1} \rangle$.

Let us consider the $2d$ Navier-Stokes equations and use the $2d$ logarithmic L^∞-estimate of Brezis & Gallouet [4, 9] in (2.11)

$$\|\mathbf{A}\|_\infty \le c_1 \|D\mathbf{A}\|_2 \left[1 + \log\left(L\frac{\|D^2\mathbf{A}\|_2}{\|D\mathbf{A}\|_2}\right)\right]^{1/2}. \tag{3.7}$$

For $\mathbf{A} \equiv D\mathbf{u}$, this has been modified into our F_n notation in [10] to give $(n \ge 3)$

$$\|D\mathbf{u}\|_\infty \le c_2 F_2^{1/2}\left[1 + \log(\lambda_0 \kappa_{n,1})\right]^{1/2}, \tag{3.8}$$

which, with $r = 1$, makes (2.11) into

$$\begin{aligned}(n-1)\dot{\kappa}_{n,1} \le\ & -\nu\kappa_{n+1,1}^3 + 2\left(c_3 F_2^{1/2}\left[1 + \log(\lambda_0\kappa_{n,1})\right]^{1/2} + \nu\lambda_0^{-2}\right)\kappa_{n,1} \\ & +\nu\kappa_{n,1}\kappa_{2,1}^2.\end{aligned} \tag{3.9}$$

The issue here is not regularity but the estimation of time averages: it has been well known (see [7, 8, 10, 26]) for many years that the $2d$ Navier-Stokes equations are regular and that a global attractor $\mathcal{A}$ is known to exist (see [7, 26]). Translated into the language of $\kappa_{n,r}$ we get estimates like $\overline{\lim}_{t\to\infty} \kappa_{n,1} \le \lambda_0^{-1}\mathcal{G}$. Use of this upper bound in both the logarithmic term and the last term in (3.9) removes them from contention in the inequalities. Hence, in terms of the $\mathcal{N}_{n,r}$ notation of (3.6), a time average of (3.9) gives

$$\left\langle \mathcal{N}_{n+1,1}^{3/2} \right\rangle \le c_5\,\nu^{-1}\lambda_0^2 \left\langle F_2^{1/2}\mathcal{N}_{n,1}^{1/2} \right\rangle \left[1 + \log\mathcal{G}\right]^{1/2} + 2\left\langle \mathcal{N}_{n,1}^{1/2} \right\rangle + c_6 \mathcal{G}\left\langle \mathcal{N}_{2,1} \right\rangle. \tag{3.10}$$

Invoking the Schwarz inequality and using the estimate for $\langle F_2 \rangle$ from Table 2, we obtain the recursion relation for $n \geq 2$

$$\langle \mathcal{N}_{n+1,1} \rangle^{3/2} \leq c_7 \, \mathcal{G}(1 + \log \mathcal{G})^{1/2} \langle \mathcal{N}_{n,1} \rangle^{1/2} + 2 \langle \mathcal{N}_{n,1} \rangle^{1/2} + c_8 \mathcal{G} \langle \mathcal{N}_{2,1} \rangle. \quad (3.11)$$

Because $\left\langle \kappa_{2,1}^2 \right\rangle \leq \lambda_0^{-2}$, we have $\langle \mathcal{N}_{2,1} \rangle \leq 1$ as our bottom rung. For $n = 2$, absorbing the last two terms in (3.11) into the constant (because $\mathcal{G} \gg 1$) we obtain

$$\langle \mathcal{N}_{3,1} \rangle = \lambda_0^2 \left\langle \kappa_{3,1}^2 \right\rangle \leq c_9 \, \mathcal{G}^{2/3}(1 + \log \mathcal{G})^{1/3}. \quad (3.12)$$

In fact, the first nontrivial member of the series of $\langle \mathcal{N}_{n,1} \rangle$ gives an estimate that coincides with the Lyapunov dimension of the attractor found in [8] which is a well known result (see Section 5). Using the recursion relation (3.11) we find that the next one is

$$\langle \mathcal{N}_{4,1} \rangle = \lambda_0^2 \left\langle \kappa_{4,1}^2 \right\rangle \leq c_{10} \, \mathcal{G}^{8/9}(1 + \log \mathcal{G})^{4/9}, \quad (3.13)$$

which rapidly asymptotes to $\mathcal{G}(1 + \log \mathcal{G})^{1/2}$ for increasing n. Both methods are dependent on $\langle F_2 \rangle$ but on no higher time average. Most of the above results can be found in [16] and are an alternative calculation of the celebrated result of Constantin, Foias & Temam [8]. Having the same estimate does not mean, however, that $\langle \mathcal{N}_{3,1} \rangle$ and the attractor dimension are formally identical in any way, but it does raise questions about the meaning of $\langle \mathcal{N}_{n,1} \rangle$ for $n \geq 4$. Foias & Prodi [13] introduced the concept of the number of determining modes of a system as far back as 1967 and, interestingly, the $c\,\mathcal{G}(1 + \log \mathcal{G})^{1/2}$ estimate coincides with the estimate found for these by Foias, Manley, Temam & Trève [12]. This has since been improved to $c\,\mathcal{G}$ by Jones & Titi [17] who discuss the inter-relation between the number of determining modes, determining nodes and the evolution of volume elements of the attractor.

Moreover, a central feature of the methods used here is that estimates are made of time averaged quantities and the role of fluctuations away from these averages may be important. Potentially, how great these could be is hard to assess. It is an open question as to whether all the estimates worse than $\mathcal{G}^{2/3}$ are a genuine reflection of potentially smaller scales in the flow caused by fluctuations, or whether they are due to a lack of sharpness in the estimates.

4   The 3d Navier-Stokes Equations

In 3d, our first task is to investigate why regularity fails to be established by the methods we have used so far. Moreover, we would also like to determine what assumptions are necessary if regularity is to be proved. Let us look firstly at the $\|D\mathbf{u}\|_\infty$ term in the F_n ladder. A calculus inequality [1, 24] yields, for each component of $\mathbf{u}$

$$\|Du_i\|_\infty \leq c \, \|D^n u_i\|_2^a \, \|u_i\|_2^{1-a}, \quad (4.1)$$

where $a = \frac{5}{2n}$ and with $2n > 5$. In consequence,

$$\|D\mathbf{u}\|_\infty \leq c \, F_n^{a/2} \, \|\mathbf{u}\|_2^{1-a}. \tag{4.2}$$

To see if we can find an absorbing ball here, the negative definite term in the ladder

$$\frac{1}{2}\dot{F}_n \leq -\nu \frac{F_n^{1+1/s}}{F_{n-s}^{1/s}} + \left(c\,\|D\mathbf{u}\|_\infty + \nu\lambda_0^{-2}\right) F_n \tag{4.3}$$

needs to be stronger than the nonlinear term. Controlling the $\|D\mathbf{u}\|_\infty$ term using (4.2), the ladder becomes

$$\frac{1}{2}\dot{F}_n \leq -\nu \frac{F_n^{1+1/s}}{F_{n-s}^{1/s}} + c\,F_n^{1+a/2}\|\mathbf{u}\|_2^{1-a} + \nu\lambda_0^{-2}F_n, \tag{4.4}$$

from which it can be seen that an absorbing ball can only be found if

$$1/s > a/2. \tag{4.5}$$

There is a degree of choice for s in the F_{n-s} term in the denominator of the Ladder Theorem. We would like to be able to go down the ladder as far as $s = n$ because F_0 is controlled for all t whereas no proof exists that F_1 is bounded in $3d$, as it is in the $2d$ case. Unfortunately, it is clear that an absorbing ball cannot be achieved this way, because equation (4.5) means that we need $s < \frac{4n}{5}$ and this cannot be fulfilled by taking $s = n$. Therefore, we can only get down to $s = n - 1$ without violating the condition for an absorbing ball. In consequence, F_1 is again the "bottom rung" of the ladder, as in the $2d$ case, and not F_0. We have no control over this bottom rung in $3d$, so there is no means of controlling the other rungs of the ladder either. It is here where the root of the problem lies. There are several different angles of attack but they all reduce to the same result in the end.

Failure to prove regularity by these methods drives us to see if we can deduce the weakest *assumption* that needs to be made to obtain regularity (see [7, 10, 20, 25, 26]). The idea is to relax the requirement that we go down to $\|\mathbf{u}\|_2^2$ in the calculus inequality in (4.1) and instead go down to L^q (for some q to be calculated) and not L^2. Hence, instead of (4.2), we write

$$\|D\mathbf{u}\|_\infty \leq c\,F_n^{a/2}\|\mathbf{u}\|_q^{1-a}, \tag{4.6}$$

where $a = \frac{2(q+3)}{(2n-3)q+6}$ with $n \geq 3$. The requirement in (4.5) that s must satisfy $as < 2$ to get an absorbing ball means that there is a restriction on q given by

$$q > \frac{3(n-2)}{n-3}. \tag{4.7}$$

However large we take n, it is clear that q must always satisfy $q > 3$. Hence in $3d$, if $\|\mathbf{u}\|_q$ ($q > 3$) is bounded for all t, then all the F_n are bounded for all

t and no singularities can occur. However, no independent proof exists that shows that $\|\mathbf{u}\|_{3+\varepsilon}$ is *a priori* bounded. The barrier between what we have proved is bounded ($\|\mathbf{u}\|_2$) and what we need to prove is bounded ($\|\mathbf{u}\|_{3+\varepsilon}$) therefore appears small, but has so far remained insurmountable.

4.1   Bounds on $\langle \kappa_{n,1} \rangle$ in 3d

Something can be said about the objects $\kappa_{n,r}$ in $3d$. No regularity proof is known so a finite time singularity in $\kappa_{n,r}$ cannot be ruled out and all the following results are valid only for weak solutions. The first step is to use (2.11) with $r = 1$ and replace $-\kappa_{n+1,1}^3$ by $-\kappa_{n,1}^3$ giving

$$(n - 1)\dot\kappa_{n,1} \leq -\nu\kappa_{n,1}^3 + \nu\kappa_{2,1}^2\kappa_{n,1} + 2\left(c_{n,1}\|D\mathbf{u}\|_\infty + \nu\lambda_0^{-2}\right)\kappa_{n,1}. \qquad (4.8)$$

$3d$ results are given in Table 3 [10] including the ones derived in this section. Some of these are necessary to prove the time average result that $\langle \kappa_{n,1} \rangle$ is bounded for all $n \geq 2$.

Lemma 3 *In 3d, weak solutions of the Navier-Stokes equations satisfy*

$$\langle \kappa_{2,1} \rangle \leq c\, L\lambda_0^{-2}\mathcal{G}^2. \qquad (4.9)$$

Proof: In $3d$ (see Appendix A) the evolution of F_1 is easily computed and is given by

$$\frac{1}{2}\dot F_1 \leq -\nu F_2 + \|\mathbf{u}\|_\infty F_1^{1/2}F_2^{1/2} + \nu\lambda_0^{-2}F_1. \qquad (4.10)$$

Using a $3d$ calculus inequality,

$$\|\mathbf{u}\|_\infty \leq c\, \|D^n\mathbf{u}\|_2^b \|D\mathbf{u}\|_2^{1-b}, \qquad (4.11)$$

3d Grashof number	$\mathcal{G} = L^{3/2}\nu^{-2}\|\mathbf{f}\|_2$
Absorbing ball for F_0	$\overline{\lim}_{t\to\infty} F_0 \leq c\, L\nu^2\mathcal{G}^2$
Time average of F_1	$\langle F_1 \rangle \leq c\, L\lambda_0^{-2}\nu^2\mathcal{G}^2$ (Leray)
Time average of $\kappa_{1,0}^2$	$\langle \kappa_{1,0}^2 \rangle \leq \lambda_0^{-2}$
Time average of $\kappa_{n,1}$	$\langle \kappa_{n,1} \rangle \leq c_{n,r}L\lambda_0^{-2}\mathcal{G}^2$
Time average of $F_n^{\frac{1}{2n-1}}$	$\left\langle F_n^{\frac{1}{2n-1}} \right\rangle \leq c_n\, \nu^{\frac{2}{2n-1}}L\lambda_0^{-2}\mathcal{G}^2$
Time average of $\|\mathbf{u}\|_\infty$	$\langle \|\mathbf{u}\|_\infty \rangle \leq c\,\nu L\lambda_0^{-2}\mathcal{G}^2$

Table 3: $3d$ results including Theorem 4, (4.22) and (4.24).

where $b = \frac{1}{2(n-1)}$ for $n \geq 2$. Taking $n = 2$ gives

$$\|\mathbf{u}\|_\infty \leq c\, F_2^{1/4} F_1^{1/4}, \tag{4.12}$$

which, when used in (4.10), gives

$$\frac{1}{2}\dot{F}_1 \leq -\frac{\nu}{4}F_2 + c\nu^{-3}F_1^3 + \nu\lambda_0^{-2}F_1. \tag{4.13}$$

This is a well known $3d$ Navier-Stokes result. Now divide by F_1^2 and time average to get

$$\left\langle \frac{F_2}{F_1^2} \right\rangle \leq c\nu^{-4}\langle F_1 \rangle \leq c\, L\lambda_0^{-2}\nu^{-2}\mathcal{G}^2. \tag{4.14}$$

where the estimate for $\langle F_1 \rangle$ comes from Table 3. There is a small bounded (F_1 is bounded from below) correction term which we ignore. Now $\langle \kappa_{2,1} \rangle$ can be written as

$$
\begin{aligned}
\langle \kappa_{2,1} \rangle &= \left\langle \left(\frac{F_2}{F_1}\right)^{1/2} \right\rangle = \left\langle \frac{F_2^{1/2}}{F_1} F_1^{1/2} \right\rangle \\
&\leq \left\langle \frac{F_2}{F_1^2} \right\rangle^{1/2} \langle F_1 \rangle^{1/2} \\
&\leq c\, L\lambda_0^{-2}\mathcal{G}^2
\end{aligned}
\tag{4.15}
$$

which is (4.9).

This result enables us to prove the following result.

Theorem 4 *For $n \geq 2$, weak solutions of the $3d$ Navier-Stokes equations satisfy*

$$\langle \kappa_{n,1} \rangle \leq c_n\, L\lambda_0^{-2}\mathcal{G}^2. \tag{4.16}$$

Remark: This result is, so far, the best we can do but the estimate is absurdly large. Its value lies in the fact that it will enable us to produce a short proof of two known results.

Proof: To handle the $\|D\mathbf{u}\|_\infty$ term we use a $3d$ calculus inequality

$$
\begin{aligned}
\|D\mathbf{u}\|_\infty &\leq c\|D^n\mathbf{u}\|_2^a \|D\mathbf{u}\|_2^{1-a} \leq c\, F_n^{a/2} F_1^{(1-a)/2} \\
&= c\,\kappa_{n,1}^{3/2} F_1^{1/2}
\end{aligned}
\tag{4.17}
$$

for $n \geq 3$ where $a = \frac{3}{2(n-1)}$. Now we divide (4.8) through by $\kappa_{n,1}^2$ and use the fact that the $\kappa_{n,1}$ are bounded below and that $\kappa_{2,1} \leq \kappa_{n,1}$ for $n \geq 2$ to get

$$\langle \kappa_{n,1} \rangle \leq c\nu^{-1} \left\langle \kappa_{n,1}^{1/2} F_1^{1/2} \right\rangle + \langle \kappa_{2,1} \rangle + \lambda_0^{-2}\kappa_{n,1}^{-1}. \tag{4.18}$$

Splitting up the first term on the right hand side of (4.18) using Hölder's inequality we find a recursion relation

$$\langle \kappa_{n,1} \rangle \leq c_n \nu^{-2} \langle F_1 \rangle + 2 \langle \kappa_{2,1} \rangle, \qquad n \geq 2, \tag{4.19}$$

where we have ignored the last term in (4.18) which is bounded and small. The recursion relation (4.19) will generate upper bounds on all the $\langle \kappa_{n,1} \rangle$ because we can control $\langle \kappa_{2,1} \rangle$ through Lemma 3.

4.2   A proof of two time average results

In [11] (see also [14]) it was shown that the quantities $\left\langle F_n^{\frac{1}{2n-1}} \right\rangle$ and $\langle \|\mathbf{u}\|_\infty \rangle$ are bounded *a priori*. Theorem 1 allows a much shorter and almost trivial proof of these. The first follows from the identity

$$\left\langle F_n^{\frac{1}{2n-1}} \right\rangle = \left\langle \kappa_{n,r}^{\frac{2(n-r)}{2n-1}} F_r^{\frac{1}{2n-1}} \right\rangle. \tag{4.20}$$

Using the Cauchy-Schwarz inequality, this becomes

$$\left\langle F_n^{\frac{1}{2n-1}} \right\rangle \leq \langle \kappa_{n,r} \rangle^{\frac{2(n-r)}{2n-1}} \left\langle F_r^{\frac{1}{2r-1}} \right\rangle^{\frac{2r-1}{2n-1}}. \tag{4.21}$$

Starting with $r = 1$ and then using the fact that both $\langle F_1 \rangle$ and $\langle \kappa_{n,1} \rangle$ are bounded above, we have

$$\left\langle F_n^{\frac{1}{2n-1}} \right\rangle \leq c_n \nu^{\frac{2}{2n-1}} L \lambda_0^{-2} \mathcal{G}^2. \tag{4.22}$$

Secondly, an inequality for $\langle \|\mathbf{u}\|_\infty \rangle$ comes about from (4.11) which can be expressed in terms of $\kappa_{n,1}$, giving

$$\|\mathbf{u}\|_\infty \leq c \, \kappa_{n,1}^{1/2} F_1^{1/2}. \tag{4.23}$$

Using the Cauchy-Schwarz inequality, we find that

$$\langle \|\mathbf{u}\|_\infty \rangle \leq c \nu L \lambda_0^{-2} \mathcal{G}^2. \tag{4.24}$$

Both these results were first found by Foias, Guillopé & Temam [11] and, in the context of the F_n-ladder, rederived in [14], but the method of calculation here is much shorter and makes plain the inter-relation between the triad of objects $\langle \kappa_{n,1} \rangle$, $\left\langle F_n^{\frac{1}{2n-1}} \right\rangle$ and $\langle \|\mathbf{u}\|_\infty \rangle$. Note also that

$$\|D\mathbf{u}\|_\infty \leq c \, \kappa_{n,1}^{3/2} F_1^{1/2}, \tag{4.25}$$

and so

$$\left\langle \|D\mathbf{u}\|_\infty^{1/2} \right\rangle \leq c \, \langle \kappa_{n,1} \rangle^{3/4} \langle F_1 \rangle^{1/4} \leq c \nu^{-3/2} \langle F_1 \rangle, \tag{4.26}$$

which is bounded.

5 The 2d Navier-Stokes Attractor and its Dimension

5.1 Attractors and their dimension

There are a variety of definitions of the dimension of a set which generalise the usual integer dimensions to fractal dimensions. These definitions have several features in common; in particular they agree for finite integer dimensional sets, and the dimension of a set is at least as large as the dimension of any of its subsets. Apart from the idea of the Hausdorff dimension, which we will not discuss here, there is the idea of the capacity or fractal dimension. The capacity dimension of a set A is defined as

$$d_C = \varlimsup_{r \searrow 0} \frac{\log M(r)}{\log \frac{1}{r}}, \tag{5.1}$$

where $M(r)$ is the minimal number of balls of radius r required to cover A.

Likewise there are a variety of definitions of the "attractor" for a dynamical system but they too have several features in common. In this article we will generally be interested in estimates or bounds on the dimension of the *global* attractor $\mathcal{A}$ defined as follows. If $\mathcal{B}$ is a compact, connected, absorbing set, absorbing all trajectories (our $\|\omega\|_2$) and $S(t)$ is the nonlinear semigroup such that $u(t) = S(t)u(0)$, then the *global attractor* $\mathcal{A}$ is defined as

$$\mathcal{A} = \cap_{t>0} S(t)\mathcal{B} \tag{5.2}$$

and has the following properties:

- $S(t)\mathcal{A} = \mathcal{A}$ forwards and backwards in time;

- for the ω-limit set of any bounded set F,

$$\omega(F) \in \mathcal{A}; \tag{5.3}$$

- $\mathcal{A}$ is compact and has the property

$$\operatorname*{dist}_{t\to\infty}\{u(t), \mathcal{A}\} = 0. \tag{5.4}$$

Thus $\mathcal{A}$ contains all the asymptotic motion for the dynamical system (see Temam [27] for a fuller explanation). It is common to talk of "multiple attractors" for a dynamical system, each of which may in its own right be considered the attractor for initial conditions within its own basin of attraction. The notion of the global attractor corresponds to the union of all possible such dynamically invariant attracting sets and more, in $\mathcal{B}$. In particular $\mathcal{A}$

contains all possible structures such as fixed points (even completely unstable points known as "repellors"), limit cycles, etc., as well as their unstable manifolds.

A connection between the system dynamics and the attractor dimension is provided by the notion of the Lyapunov exponents via the Kaplan-Yorke formula. Roughly speaking, the Lyapunov exponents control the exponential growth or contraction of volume elements in phase space and the Kaplan-Yorke formula expresses the balance between volume growth and contraction realised on the attractor. The (global) Lyapunov exponents μ_i are determined according to the rule that the sum of the first n exponents gives the (largest possible) asymptotic exponential growth rate of n-volumes. If the sum of the first n exponents is negative, then all n-volumes decay to zero exponentially. Before giving the precise definition of the global Lyapunov exponents though, we give an indication of their relation to the attractor dimension through the Kaplan-Yorke formula.

The connection between the Lyapunov exponents and the capacity (fractal) dimension of the attractor is heuristically established by the following argument. First suppose that the sum of the first N exponents is negative, but that the sum of the first $N-1$ is not. Then the attractor dimension should be less than N because it cannot contain any N-dimensional subsets, and we presume that we may cover the attractor with N-dimensional balls of radius r. Suppose that it requires $M(r)$ such balls to cover the attractor. Then we let the system dynamics operate for a time t and observe that the N-dimensional balls of volume $\sim r^N$ evolve into N-dimensional ellipsoids of volume $\sim r^N \exp\{[\mu_1 + \ldots + \mu_N]t\}$ which still cover the attractor. The smallest axis of the ellipsoids is $r\exp(\mu_N t) < r$, because μ_N is necessarily the most negative exponent. Then we cover each ellipsoid with balls of radius $r\exp(\mu_N t)$, which requires $\sim \exp\{[\mu_1 + \ldots + \mu_{N-1} + (1-N)\mu_N]t\}$ smaller balls. So the total number of smaller balls required to cover the attractor is

$$M\left(r\exp(\mu_N t)\right) \sim \exp\left\{[\mu_1 + \ldots + \mu_{N-1} - (N-1)\mu_N]t\right\} M(r). \qquad (5.5)$$

According to formula (5.1) for the capacity dimension, then

$$\begin{aligned}
d_C &\approx \varlimsup_{t\to\infty} \frac{\log\left[M\left(re^{\mu_N t}\right)\right]}{\log\left[\frac{1}{r}e^{-\mu_N t}\right]} \\
&= N - 1 + \frac{\mu_1 + \ldots + \mu_{N-1}}{-\mu_N}.
\end{aligned} \qquad (5.6)$$

This is the Kaplan-Yorke formula. Note that according to the definition of N, the ratio of exponents in (5.6) satisfies

$$0 \le \frac{\mu_1 + \ldots + \mu_{N-1}}{-\mu_N} < 1, \qquad (5.7)$$

so the Kaplan-Yorke formula generally yields a noninteger dimension between $N - 1$ and N. We use it to define the *Lyapunov* dimension of the attractor:

$$d_L = N - 1 + \frac{\mu_1 + \ldots + \mu_{N-1}}{-\mu_N}, \tag{5.8}$$

where

$$\sum_{n=1}^{N-1} \mu_n \geq 0 \qquad \text{but} \qquad \sum_{n=1}^{N} \mu_n < 0 \tag{5.9}$$

and $N - 1 < d_L < N$. The Lyapunov dimension differs conceptually from either the capacity or the Hausdorff dimensions in that its definition relies explicitly on the dynamics generating the attractor.

5.2   The 2d attractor dimension estimate

In a seminal paper, Constantin & Foias [6] developed the application of global Lyapunov exponents for the Navier Stokes equations and other dissipative PDE's (see also [7, 10, 27]). To actually do this for the 2d Navier-Stokes equations, we consider the vorticity evolution (1.4) to be the defining equation, and recall that there are periodic boundary conditions on the square domain $\Omega \equiv [0, L]^2$. The velocity vector field is expressed in terms of the vorticity as

$$\mathbf{u} = \left(-\frac{\partial \Delta^{-1} \omega}{\partial y}, \frac{\partial \Delta^{-1} \omega}{\partial x} \right), \tag{5.10}$$

where, without loss of generality, we may assume that the spatially averaged vorticity vanishes so that the Laplacian may be appropriately inverted.

Upper bounds on the attractor dimension for the system are determined by considering the time evolution of volume elements in the system's configuration space which, in this case, is $L^2(\Omega)$. If, say, all N-dimensional volumes contract to zero volume as $t \to \infty$, then the attractor cannot contain any N-dimensional subsets and hence $d_{att} \leq N$. The goal is to determine the smallest possible N with this property, as it constitutes an upper bound on the attractor's dimension. This value of N is the equivalent to the N in equation (5.9).

We restrict our consideration to infinitesimal volume elements whose evolutions are controlled by the linearised Navier-Stokes equations, linearised about an arbitrary solution on the attractor. For (1.4), the linearised equation for the difference $\delta\omega$ between two neighbouring solutions is

$$\frac{\partial \delta\omega}{\partial t} = -\mathbf{A}(t)\delta\omega = -\mathbf{u} \cdot \nabla \delta\omega - \delta\mathbf{u} \cdot \nabla \omega + \nu \Delta \delta\omega, \tag{5.11}$$

where the associated variation in the velocity is

$$\delta\mathbf{u} = \left(-\frac{\partial \Delta^{-1} \delta\omega}{\partial y}, \frac{\partial \Delta^{-1} \delta\omega}{\partial x} \right). \tag{5.12}$$

The magnitude of the infinitesimal N-dimensional volume element spanned by $\delta\omega_1(t), \ldots, \delta\omega_N(t)$ is

$$V_N(t) = |\delta\omega_1(t) \wedge \ldots \wedge \delta\omega_N(t)|, \tag{5.13}$$

each edge of which develops according to (5.11), itself evolving according to

$$V_N(t) = V_N(0) \exp\left(-\int_0^t Tr[\mathbf{A}(t')\mathbf{P}_N(t')]dt'\right). \tag{5.14}$$

In the above, $\mathbf{P}_N(t)$ is the projection onto the linear subspace of $L^2(\Omega)$ spanned by $\delta\omega_1(t), \ldots, \delta\omega_N(t)$. In order for N to be an upper bound on the attractor dimension, the volume elements $V_N(t)$ about *any* solution $\omega(t)$ on the attractor must vanish as $t \to \infty$. Rewriting (5.14),

$$\begin{aligned} V_N(t) &= V_N(0) \exp\left[-t\left(\frac{1}{t}\int_0^t Tr[\mathbf{A}(t')\mathbf{P}_N(t')]dt'\right)\right] \\ &\xrightarrow[t\to\infty]{} V_N(0) \exp\left\{-t\langle Tr[\mathbf{A}\mathbf{P}_N]\rangle\right\}, \end{aligned} \tag{5.15}$$

where $\langle\cdot\rangle$ denotes the largest possible time average. Thus, to determine an upper bound on the attractor dimension we look for the smallest N which satisfies

$$\langle -Tr[\mathbf{A}\mathbf{P}_N]\rangle < 0 \tag{5.16}$$

for all solutions $\omega(t)$. Let $\{\phi_i(t)\}$ be orthonormal basis functions spanning $\mathbf{P}_N L^2[\Omega]$ for $1 \le i \le N$, and let the associated vector fields $\mathbf{v}_1(t), \ldots, \mathbf{v}_N(t)$ be

$$\mathbf{v}_n = \left(-\frac{\partial\Delta^{-1}\phi_n}{\partial y}, \frac{\partial\Delta^{-1}\phi_n}{\partial x}\right). \tag{5.17}$$

Estimate the trace as follows:

$$\begin{aligned} Tr[\mathbf{A}(t)\mathbf{P}_N(t)] &= \sum_{n=1}^{N} \int_\Omega \phi_n(t)\mathbf{A}(t)\phi_n(t)d^2x \\ &= \nu\sum_{n=1}^{N}\int_\Omega |\nabla\phi_n|^2\,d^2x + \sum_{n=1}^{N}\int_\Omega \phi_n(\mathbf{u}\cdot\nabla\phi_n + \mathbf{v}_n\cdot\nabla\omega)d^2x \\ &= \nu Tr[-\Delta\mathbf{P}_N] + \sum_{n=1}^{N}\int_\Omega \phi_n\mathbf{v}_n\cdot\nabla\omega d^2x, \end{aligned} \tag{5.18}$$

where we have used the fact that $\mathbf{u}$ is divergence free to eliminate one of the terms. Because we know the spectrum of the Laplacian explicitly, the real work consists of finding good sharp upper bounds on the last sum in (5.18).

Using the Schwarz inequality, we have

$$\left|\sum_{n=1}^{N}\int_\Omega \phi_n\mathbf{v}_n\cdot\nabla\omega d^2x\right| \le \int_\Omega \left[\left(\sum_{n=1}^{N}|\phi_n|^2\right)^{1/2}\left(\sum_{n=1}^{N}|\mathbf{v}_n|^2\right)^{1/2}|\nabla\omega|\right]d^2x. \tag{5.19}$$

Using one of Hölder's inequalities, we pull out the sum of the squares of the $\mathbf{v}_n$'s in the L^∞ norm, to find

$$\left| \sum_{n=1}^{N} \int_\Omega \phi_n \mathbf{v}_n \cdot \nabla \omega d^2 x \right| \leq \left(\sum_{n=1}^{N} \|\mathbf{v}_n\|_\infty^2 \right)^{1/2} \int_\Omega \left[\left(\sum_{n=1}^{N} \phi_n^2 \right)^{1/2} |\nabla \omega| \right] d^2 x. \quad (5.20)$$

The Cauchy-Schwarz inequality is now used to separate the two factors inside the integral above:

$$\left| \sum_{n=1}^{N} \int_\Omega \phi_n \mathbf{v}_n \cdot \nabla \omega d^2 x \right| \leq \left(\sum_{n=1}^{N} \|\mathbf{v}_n\|_\infty^2 \right)^{1/2} \left(\int_\Omega \sum_{n=1}^{N} \phi_n^2 d^2 x \right)^{1/2} \|\nabla \omega\|_2. \quad (5.21)$$

To estimate the first factor above we now use Constantin's theorem [5] which provides L^∞ estimates on collections of functions whose *gradients* are orthonormal. We use it in the form where it is applicable to the sum of the squares of the $\mathbf{v}_n$'s

Theorem 5 *If, for $1 \leq n \leq N$, $\mathbf{v}_n$ are functions whose gradients are orthonormal, that is $\int_\Omega (\nabla v_{n\beta}) \cdot (\nabla v_{m\beta}) \, d^2 x = \delta_{mn}$, then*

$$\left(\sum_{n=1}^{N} \|\mathbf{v}_n\|_\infty^2 \right) \leq c \left\{ 1 + \log(L^2 Tr[-\Delta \mathbf{P}_N]) \right\}, \quad (5.22)$$

where

$$Tr[-\Delta \mathbf{P}_N] = \sum_{n=1}^{N} \int_\Omega |\nabla \phi_n|^2 \, d^2 x \quad (5.23)$$

and the constant c is independent of N.

The proof is given in [10]. This theorem was first proved by Constantin [5], but in the form given above was proved in [9].

Continuing with our estimation of (5.21) above, its middle factor is evaluated by recalling that the ϕ_n's are orthonormal functions so that

$$\sum_{n=1}^{N} \int_\Omega \phi_n^2 \, d^2 x = N. \quad (5.24)$$

The first term in (5.18), the trace of the Laplacian in an N-dimensional subspace, is easily estimated

$$Tr[-\Delta \mathbf{P}_N] \geq c \, N^2 L^{-2}. \quad (5.25)$$

Hence we rewrite (5.24) as

$$\int_\Omega \sum_{n=1}^{N} \phi_n^2 \, d^2 x \leq c \, (Tr[-\Delta \mathbf{P}_N])^{1/2} L. \quad (5.26)$$

Putting together equations (5.21), (5.22) and (5.26) we arrive at

$$\left| \sum_{n=1}^{N} \int_{\Omega} \phi_n \mathbf{v}_n \cdot \nabla \omega \, d^2 x \right| \leq c \|\nabla \omega\|_2 \left[g \left(L^2 Tr[-\Delta \mathbf{P}_N] \right) \right]^{1/2}, \qquad (5.27)$$

where $g(\zeta) = \sqrt{\zeta}(1 + \log \zeta)$. Taking the time average we have

$$\left\langle \left| \sum_{n-1}^{N} \int_{\Omega} \phi_n \mathbf{v}_n \cdot \nabla \omega \, d^2 x \right| \right\rangle \leq c \left\langle \|\nabla \omega\|_2 \left[g \left(L^2 Tr[-\Delta \mathbf{P}_N] \right) \right]^{1/2} \right\rangle. \qquad (5.28)$$

Using the Cauchy-Schwarz inequality on the time average on the right-hand side above gives

$$\left\langle \left| \sum_{n=1}^{N} \int_{\Omega} \phi_n \mathbf{v}_n \cdot \nabla \omega \, d^2 x \right| \right\rangle \leq c \left\langle \|\nabla \omega\|_2^2 \right\rangle^{1/2} \left\langle g \left(L^2 Tr[-\Delta \mathbf{P}_N] \right) \right\rangle^{1/2}. \qquad (5.29)$$

Now, the function $g(\zeta)$ is concave for $\zeta > 1/e$. For large values of N which are appropriate for turbulent flows, $L^2 Tr[-\Delta \mathbf{P}_N] \gg 1$, and we invoke Jensen's inequality ($\langle g(\zeta) \rangle \leq g(\langle \zeta \rangle)$ for g concave) to find

$$\left\langle \left| \sum_{n=1}^{N} \int_{\Omega} \phi_n \mathbf{v}_n \cdot \nabla \omega d^2 x \right| \right\rangle \leq c \left\langle \|\nabla \omega\|_2^2 \right\rangle^{1/2} \left[g \left(\left\langle L^2 Tr[-\Delta \mathbf{P}_N] \right\rangle \right) \right]^{1/2}. \qquad (5.30)$$

Therefore, the time averaged trace in (5.15) and (5.16), controlling the exponential growth or contraction of volume elements, is estimated by

$$\langle Tr[\mathbf{AP}_N] \rangle \geq \frac{\nu}{L^2} \left\langle L^2 Tr[-\Delta \mathbf{P}_N] \right\rangle$$
$$- c \left\langle \|\nabla \omega\|_2^2 \right\rangle^{1/2} \left[g \left(\left\langle L^2 Tr[-\Delta \mathbf{P}_N] \right\rangle \right) \right]^{1/2}. \qquad (5.31)$$

Next we use the $2d$ Grashof number $\mathcal{G}$, defined already in Section 1. Multiplying the vorticity version of the Navier-Stokes equation (1.4) by ω, integrating over the spatial variables and taking the time average, we have

$$\nu \left\langle \|\nabla \omega\|_2^2 \right\rangle = \left\langle \int_{\Omega} \omega \hat{\mathbf{k}} \cdot \text{curl} \, \mathbf{f} \, d^2 x \right\rangle. \qquad (5.32)$$

A simple integration by parts and application of Cauchy's inequality yields

$$\nu \left\langle \|\nabla \omega\|_2^2 \right\rangle \leq \left\langle \|\nabla \omega\|_2 \right\rangle \|\mathbf{f}\|_2 \leq \left\langle \|\nabla \omega\|_2^2 \right\rangle^{1/2} \|\mathbf{f}\|_2, \qquad (5.33)$$

so that

$$\left\langle \|\nabla \omega\|_2^2 \right\rangle^{1/2} \leq \|\mathbf{f}\|_2/\nu = \nu \mathcal{G}/L^2. \qquad (5.34)$$

Define $\mathcal{N}$ by

$$\mathcal{N}^2 = L^2 \left\langle Tr[-\Delta \mathbf{P}_N] \right\rangle. \qquad (5.35)$$

Then the trace formula in (5.31) is

$$\langle Tr[\mathbf{A}\mathbf{P}_N]\rangle \geq \frac{\nu}{L^2}\left[\mathcal{N}^2 - c\mathcal{G}\,\mathcal{N}^{1/2}(1 + \log\mathcal{N})^{1/2}\right]. \tag{5.36}$$

As N increases, then so does $\mathcal{N}$ and for a given $\mathcal{G}$ it eventually forces the right-hand side of (5.36) to become positive so that all volume elements of dimension higher than N contract to zero. If the logarithmic term was absent in (5.36), then this crossover point would occur when $N \sim \mathcal{N} \sim \mathcal{G}^{2/3}$. The logarithm introduces corrections to this. The precise answer, including the logarithmic correction, leads us to conclude that all N-dimensional volume elements contract to zero when

$$N \geq c\mathcal{G}^{2/3}(1 + \log\mathcal{G})^{1/3}, \tag{5.37}$$

which is the same as (3.12). To express d_{att} in terms of the Kraichnan length λ_{K_r} defined in (1.18) instead of $\mathcal{G}$, we return to (5.31) and express $\langle\|\nabla\omega\|_2^2\rangle$ in terms of the average enstrophy dissipation rate χ given by (1.17). We finally obtain

$$d_{att} \leq c\left(\frac{L}{\lambda_{K_r}}\right)^2\left[1 + \log\left(\frac{L}{\lambda_{K_r}}\right)\right]^{1/3}. \tag{5.38}$$

A   Estimates for the Navier-Stokes Equations

While the ladder given in Table 1 is the most general result for arbitrary n, for the three lowest rungs, there are more sensitive estimates for F_0, F_1 and F_2. Having good bounds on these is essential because the ladder acts as a form of a recurrence relation.

A.1   Estimates for F_0

For a divergence free flow ($\nabla \cdot \mathbf{u} = 0$), let us again write down the Navier-Stokes equations

$$\mathbf{u}_t + \mathbf{u} \cdot \nabla\mathbf{u} = \nu\Delta\mathbf{u} - \nabla p + \mathbf{f}, \tag{A.1}$$

with $\nabla \cdot \mathbf{u} = 0$. The only level where it is not necessary to use our standard F_n-notation is at $n = 0$. Using a well known vector identity, the Navier-Stokes equations in (A.1) can be rewritten as

$$\mathbf{u}_t + \boldsymbol{\omega} \times \mathbf{u} = \nu\Delta\mathbf{u} - \nabla(p + \frac{1}{2}u^2) + \mathbf{f}. \tag{A.2}$$

Taking the scalar product of $\mathbf{u}$ with (A.2) and integrating over the domain gives the evolution of the energy $\frac{1}{2}\|\mathbf{u}\|_2^2$. The nonlinear term immediately vanishes and the energy evolves according to

$$\frac{d}{dt}\left(\frac{1}{2}\|\mathbf{u}\|_2^2\right) = -\nu\|\nabla\mathbf{u}\|_2^2 + \int_\Omega \mathbf{u} \cdot \mathbf{f}\, d^d x. \tag{A.3}$$

An appeal to Poincaré's inequality, which is used on the Laplacian term, and use of the Cauchy-Schwarz inequality on the forcing term, gives

$$\frac{d}{dt}\left(\frac{1}{2}\|\mathbf{u}\|_2^2\right) \leq -\nu k_1^2\|\mathbf{u}\|_2^2 + \|\mathbf{u}\|_2\|\mathbf{f}\|_2, \tag{A.4}$$

where $k_1 = 2\pi/L$. To express the $\|\mathbf{u}\|_2^2$ in terms of the forcing it is easier if we use the dimensionless Grashof numbers defined in (1.11), given by

$$2d: \quad \mathcal{G} = \frac{L^2\|\mathbf{f}\|_2}{\nu^2}, \qquad\qquad 3d: \quad \mathcal{G} = \frac{L^{3/2}\|\mathbf{f}\|_2}{\nu^2}. \tag{A.5}$$

Then, from (A.4), limsup estimates for $\|\mathbf{u}\|_2^2$ are found using Gronwall's Lemma as

$$2d: \quad \varlimsup_{t\to\infty}\|\mathbf{u}\|_2^2 \leq c\nu^2\mathcal{G}^2, \qquad\qquad 3d: \quad \varlimsup_{t\to\infty}\|\mathbf{u}\|_2^2 \leq cL\nu^2\mathcal{G}^2, \tag{A.6}$$

where the dimensionless constants are denoted by c. The energy estimates given in (A.6) above, while useful, are only part of what we need in order to find an estimate for F_0. Because all the F_n contain forcing terms, estimates for these must also be included. From (2.4), we recall that

$$F_0 = \|\mathbf{u}\|_2^2 + \nu^{-2}L^4\|\mathbf{f}\|_2^2. \tag{A.7}$$

It is desirable to express the extra term $\nu^{-2}L^4\|\mathbf{f}\|_2^2$ in terms of the Grashof numbers defined above. From (2.2) we recall that our choice of forcing function is such that it possesses a smallest scale (see (2.3)), called λ_f, which appears in the definition of the standard length scale λ_0 (see (2.5))

$$\lambda_0^{-2} = L^{-2} + \lambda_f^{-2}. \tag{A.8}$$

When the forcing is included, (A.6) is modified to become

$$2d: \quad \varlimsup_{t\to\infty} F_0 \leq c\nu^2\mathcal{G}^2, \qquad\qquad 3d: \quad \varlimsup_{t\to\infty} F_0 \leq cL\nu^2\mathcal{G}^2. \tag{A.9}$$

These results are listed in Tables 2 and 3.

A.2   Estimates for $\langle F_1 \rangle$ and $\langle \kappa_{1,0}^2 \rangle$

The ladder for $n = 0$ is

$$\frac{1}{2}\dot{F}_0 \leq -\nu F_1 + \nu\lambda_0^{-2}F_0. \tag{A.10}$$

Time averaging (A.10) and using (A.6) therefore produces

$$2d: \quad \langle F_1 \rangle \leq c\lambda_0^{-2}\nu^2\mathcal{G}^2, \qquad\qquad 3d: \quad \langle F_1 \rangle \leq cL\lambda_0^{-2}\nu^2\mathcal{G}^2. \tag{A.11}$$

The $3d$ estimate above, in particular, is the time averaged version of Leray's inequality (see (1.13)). Note that our estimate for F_1 in (A.11) includes λ_0^{-2} and thereby takes account of spectral information of the forcing. To obtain a version of Leray's inequality in terms of $\kappa_{n,r}^2$, we can also easily see that dividing (A.10) by F_0 and time averaging gives

$$\left\langle \kappa_{1,0}^2 \right\rangle \leq \lambda_0^{-2}, \tag{A.12}$$

which is true for both $2d$ and $3d$. Again, these results are listed in Tables 2 and 3.

A.3 Estimates for $\overline{\lim}_{t\to\infty} F_1$, $\langle F_2 \rangle$ and $\left\langle \kappa_{2,1}^2 \right\rangle$

Now we turn to the evolution of the enstrophy $\int_\Omega |\omega|^2 \, d^d x$ which is given by

$$\frac{1}{2}\frac{d}{dt}\int_\Omega |\omega|^2 \, d^d x = -\nu\int_\Omega |\nabla\omega|^2 \, d^d x - \int_\Omega \omega\cdot[\mathbf{u}\cdot\nabla\omega]\, d^d x$$
$$+ \int_\Omega \omega\cdot[\omega\cdot\nabla\mathbf{u}]\, d^d x + \int_\Omega \omega\cdot\operatorname{curl}\mathbf{f}\, d^d x. \tag{A.13}$$

Because of the periodicity of the boundary conditions and the fact that $\nabla\cdot\mathbf{u} = 0$, the integral $\int_\Omega \omega\cdot[\mathbf{u}\cdot\nabla\omega]\, d^d x$ vanishes for both $2d$ and $3d$. In $2d$, $\omega\cdot\nabla\mathbf{u} = 0$ because the vorticity vector ω is perpendicular to a $2d$ flow in the $x-y$ plane. This enables us to easily find an absorbing ball for the $2d$ enstrophy. In terms of F_1, we find

$$\frac{1}{2}\dot{F_1} \leq -\nu F_2 + \nu\lambda_0^{-2} F_1. \tag{A.14}$$

Using integration by parts and the Schwarz inequality we know that

$$F_1^2 \leq F_0 F_2, \tag{A.15}$$

and so, using (A.9), we find that

$$\overline{\lim_{t\to\infty}} F_1 \leq c\,\lambda_0^{-2}\nu^2\mathcal{G}^2. \tag{A.16}$$

Considering (A.14) again, we may time average the equation to obtain $\langle F_2 \rangle$ or we can divide by F_1 and then time average to find $\left\langle \kappa_{2,1}^2 \right\rangle$.

References

[1] R.A. Adams. *Sobolev Spaces*. Academic Press, NY, 1975.

[2] M. Bartuccelli, C. Doering, and J.D. Gibbon. Ladder theorems for the $2d$ and $3d$ Navier-Stokes equations on a finite periodic domain. *Nonlinearity* **4**, 531-542, 1991.

[3] M. Bartuccelli, C.R. Doering, J.D. Gibbon and S.J.A. Malham. Length scales in solutions of the Navier-Stokes equations. *Nonlinearity* **6**, 549-568, 1993.

[4] H. Brezis and T. Gallouet. Nonlinear Schrödinger evolution equations. *Nonlin. Anal. Thy Meth. Appl.* **4**, 677-681, 1980.

[5] P. Constantin. Collective L^∞ estimates for families of functions with orthonormal derivatives. *Indiana Univ. Math. J.* **36**, 603-615, 1987.

[6] P. Constantin and C. Foias. Global Lyapunov exponents, Kaplan-Yorke formulas and the dimension of the attractors for $2d$ Navier-Stokes equations. *Comm. Pure Appl. Math.* **38**, 1-27, 1985.

[7] P. Constantin and C. Foias. *Navier-Stokes Equations*. University of Chicago Press, 1988.

[8] P. Constantin, C. Foias and R. Temam. On the dimension of the attractors in two-dimensional turbulence. *Physica D* **30**, 284-296, 1988.

[9] C.R. Doering and J.D. Gibbon. A note on the Constantin-Foias-Temam attractor dimension estimate for two-dimensional turbulence. *Physica D* **48**, 471-480, 1991.

[10] C.R. Doering and J.D. Gibbon. *Applied Analysis of the Navier-Stokes Equations*. CUP, Cambridge, 1995.

[11] C. Foias, C. Guillopé and R. Temam. New a priori estimates for Navier-Stokes equations in dimension 3. *Comm. in Partial Diff. Equat.* **6**, 329-359, 1981.

[12] C. Foias, O.P. Manley, R. Temam and Y.M. Trève. Asymptotic analysis of the Navier-Stokes equations. *Physica D* **9**, 157-188, 1983.

[13] C. Foias and G. Prodi. Sur le comportement global des solutions non stationnaires des èquations de Navier-Stokes en dimension deux. *Rend. Sem. Mat. Univ. Padova* **39**, 1-34, 1967.

[14] J.D. Gibbon. Derivation of $3d$ Navier Stokes length scales from a result of Foias, Guillopé and Temam. *Nonlinearity* **7**, 245-252, 1994.

[15] J.D. Gibbon. Length scales and ladder theorems for $2d$ and $3d$ convection. *Nonlinearity* **8**, 81-92, 1995.

[16] J.D. Gibbon. A voyage around the Navier-Stokes equations. Submitted to *Physica D*, 1995.

[17] D.A. Jones and E.S. Titi. Upper bounds for the number of determining modes, nodes and volume elements for the Navier-Stokes equations. *Indiana Univ. Math. J.* **42**, 875-887, 1993.

[18] A.N. Kolmogorov. Local structure of turbulence in an incompressible fluid at very high Reynolds numbers. *Dokl. Akad. Nauk. SSSR* **30**, 299-303, 1941.

[19] R.H. Kraichnan and D. Montgomery. Two-dimensional turbulence. *Rep. Prog. Phys.* **43**, 547-619, 1980.

[20] O.A. Ladyzhenskaya. *The Mathematical Theory of Viscous Incompressible Flow*. Gordon and Breach, New York, second edition, 1963.

[21] J. Leray. Essai sur le mouvement d'un liquide visquex emplissant l'espace. *Acta Math.* **63**, 193-248, 1934.

[22] A. Majda. Vorticity and the mathematical theory of incompressible fluid flow. *Comm. Pure Appl. Math.* **39**, 187-220, 1986.

[23] A. Majda. Vorticity, turbulence and acoustics in fluid flow. *SIAM Rev.* **33**, 349-388, 1991.

[24] L. Nirenberg. On elliptic partial differential equations. *Annali della Scuola Norm. Sup.* **13**, 115-162, 1959.

[25] J. Serrin. On the Interior Regularity of Weak Solutions of the Navier-Stokes Equations. *Arch. Rat. Mech. Anal.* **9**, 187-195, 1962.

[26] R. Temam. *The Navier-Stokes Equations and Non-linear Functional Analysis*. CBMS-NSF Regional Conference Series in Applied Mathematics, SIAM, 1983.

[27] R. Temam. *Infinite Dimensional Dynamical Systems in Mechanics and Physics*, Vol 68 of *Applied Mathematical Sciences*, Springer-Verlag, New York, 1988.

Organised Chaos in Fluid Dynamics

T. Mullin and J.J. Kobine
Department of Atmospheric, Oceanic and Planetary Physics,
Clarendon Laboratory,
University of Oxford,
Oxford

1   Introduction

Obtaining an understanding of any physical phenomenon will inevitably be
achieved through a combination of experimental and theoretical investigations. A typical scenario is that an experiment generates data which can then
be compared with the predictions of a theoretical model. On the strength of
this comparison, both the experiment and the model may be refined until
a stage is reached when the model captures the essential behaviour of the
system under investigation. However, it is only on very rare occasions that
the actual equations that govern the evolution of a physical process can be
derived explicitly. If it is possible to write down the governing equations,
then the problem is generally considered to be solved at that point.

The equations that govern the flow of a fluid, the *Navier–Stokes equations*,
have been known since the middle of the nineteenth century. They are in fact
just the mathematical statement of Newton's Second Law of Motion applied
to a continuous medium. Together with appropriate boundary conditions and
a statement of mass conservation, they are believed to specify completely the
dynamical state of a fluid. However, despite more than a century of scientific
investigation, relatively few connections have ever been made between the
Navier–Stokes equations themselves and the panoply of behaviour exhibited
by fluids in motion. At best, exact analytical solutions can be found for some
extremely simple idealised cases of steady two-dimensional flow in highly
symmetrical domains. More generally, for the vast majority of flows that are
encountered in nature, the governing equations might as well not exist, since
we are unable to make progress towards closed-form solutions.

The difficulties arise from the fact that the Navier–Stokes equations are a
set of nonlinear coupled partial differential equations. As such, they possess
an infinite number of degrees of freedom and must be solved with respect to
space as well as time. Indeed, the greatest challenge in classical physics is presented by the phenomenon of *turbulence*, where the velocity field is spatially
disordered and temporally irregular. There are no analytical functions that
can describe such a vector field, so even if we knew how to obtain a solution,
we simply do not have the ability to express that solution mathematically.

147

Given such an impasse, there are essentially two conventional courses of action. One can proceed with experimental investigations, while at the same time developing empirical models that fit the observed behaviour in particular situations. Such an approach has the potential for providing practical solutions to engineering problems, where typically it is bulk quantities such as flow rate or drag coefficient that are sought. Alternatively, one can make use of increasingly powerful computational resources to generate numerical representations of flow fields from discretised versions of the governing equations. However, despite their respective merits, empiricism and simulation can never be more than pragmatic approaches that fall short of tackling the fundamental aspects of the physics of fluid motion.

In recent years, however, an alternative perspective has emerged that is providing genuine insights into the mechanisms behind some complicated fluid behaviour. This approach is based on ideas from the mathematical disciplines of bifurcation theory and finite-dimensional dynamical systems. Perhaps the most celebrated example of this is to be found in the work of Lorenz [30] who showed that practical unpredictability is a property of a set of three coupled ordinary differential equations with simple quadratic nonlinearities. The model is a severely truncated set of equations which portrays some aspects of fluid convection. Nowadays, it is often used as an example of a low-dimensional system that exhibits temporal chaos i.e. a dynamical motion on an attractor such that nearby trajectories diverge from each other exponentially on average, the so called 'sensitivity to initial conditions' (Ruelle [47]). The attractor of the Lorenz system in its chaotic state is an example of a 'strange attractor'. These mathematical entities were proposed by Ruelle & Takens [48] to lie at the heart of an explanation of fluid turbulence. A lively discussion of the advantages and disadvantages of this approach compared with the pragmatic approaches described above can be found in Lumley [31].

In the laboratory, it has been found that certain flows undergo a well-defined sequence of transitions from initially steady motion to flow that is temporally aperiodic (see Swinney & Gollub [53] for a review of this work). At each transition, a flow that was originally stable to infinitesimal perturbations becomes unstable and therefore is no longer sustainable in practice. Stability is then taken up by a new flow configuration, and this flow remains qualitatively the same until the next instability is encountered. The remarkable feature of this process is that, despite the potentially infinite number of degrees of freedom, the dynamics of the flow during these various transitions only ever involves a small number of temporal modes.

The flow of a fluid between concentric rotating cylinders, commonly known as *Taylor–Couette flow*, is an example of a fluid mechanical system that exhibits a bifurcation sequence to low-dimensional chaos. This system is shown schematically in Fig. 1. A review of the first sixty years of research carried out since the pioneering work of G.I. Taylor in 1923 is given by DiPrima & Swinney [18]. Since it was first considered, there has been considerable research

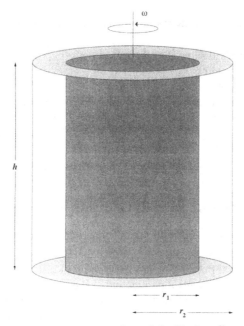

Figure 1: Schematic representation of the Taylor–Couette geometry.

effort made to study all possible realisations of the basic Taylor–Couette geometry. From here on we will only consider the case where the inner cylinder rotates and the outer one is at rest. Also, we will not attempt to give an extensive review of all the modern research on the problem, but instead will focus on those points which we believe to be pertinent to our discussion of structured low-dimensional chaos.

Interest in this aspect of the problem began when Gollub & Swinney [23] demonstrated the appearance of chaos as a result of a sequence of bifurcations as the rotation rate of the inner cylinder was increased. Until that time, observation of the flow was mainly carried out using visualisation techniques which showed the sequence of featureless shear flow, Taylor cells, waves and then disordered motion. An important aspect of Gollub and Swinney's work was to provide evidence which supported the ideas of Ruelle & Takens [48] concerning a finite bifurcation sequence to turbulence in contrast to the earlier model of Landau, who suggested a continuous transition. Since that time, observations of other 'routes to chaos' have been recorded in the Taylor–Couette problem, with those by Pfister [43], Brandstater & Swinney [6], Mullin & Price [40] and Price & Mullin [46] being good examples. Thus it is misleading to think that there is a single universal route to chaos even within this apparently simple flow geometry. One of the fascinations of the problem is that it can exhibit a plethora of dynamical processes when investigated in detail. A recent review of some of this work is given by Tagg [54].

In this article we will show that there is organisation of observed chaos by multiple bifurcation points in the solution set. Thus the plethora of apparently different types of dynamical chaos may be thought of as parts of a grand global whole. We will then review some experimental tests of the structural stability of these features where the effects of symmetries and connectedness of the fluid domain are systematically varied. Finally, we will discuss the application of these ideas to the important practical field of 'open flows' where both temporal and spatial effects are thought to be important.

2 Codimension and Organising Centres

It is generally accepted that finite-dimensional chaos will often, although not exclusively, arise in physical systems as the end result of a bifurcation sequence. The critical points that form this chain of instabilities are encountered by varying certain control parameters of the system. By measuring the critical values at which bifurcations occur, it is possible to map out paths or loci of critical points in the parameter space of the problem under consideration.

An important concept that must be introduced at this stage is the *codimension* of a bifurcation, which is defined here as the smallest dimension of parameter space that contains the bifurcation in a persistent way. This is the definition adopted by Guckenheimer & Holmes [25], amongst others, but it is essential to note that there is an alternative definition in use in singularity theory as given by Golubitsky & Schaeffer [24]. Simple bifurcations that are structurally stable, such as the saddle-node and Hopf bifurcations (Fig. 2), have a codimension of one since it only requires variation of one parameter to encounter the bifurcation. However, a pitchfork bifurcation (Fig. 3) is structurally unstable to two different types of asymmetric perturbations and so has a codimension of three since control must be exercised over both the bifurcation parameter and the two parameters controlling the asymmetry for this bifurcation to be encountered (see Golubitsky & Schaeffer [24]).

A more complicated type of codimension-2 bifurcation is also possible. For example, consider the case where a system is controlled by two independent parameters. Variation of one of the parameters will typically lead to a distinct sequence of instabilities involving codimension-1 bifurcations. If the second parameter is also varied, paths of bifurcation points can be traced out in the two-dimensional parameter space. It is now possible for two loci of codimension-1 bifurcations to intersect so that there is a double bifurcation, or *codimension-2 point*, where two qualitatively different types of instability occur simultaneously. It is known that such codimension-2 points can act as *organising centres* for the global dynamics of systems governed by ordinary

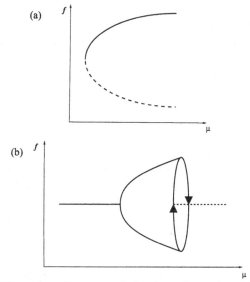

Figure 2: Bifurcation diagrams of simple codimension-1 bifurcations: (a) saddle node; (b) Hopf. Solid/dashed lines denote stable/unstable solution branches respectively. f is a linear functional that distinguishes between solutions and μ is the bifurcation parameter.

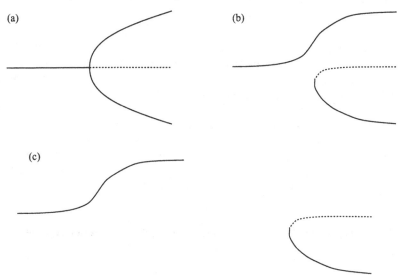

Figure 3: Bifurcation diagrams of pitchfork bifurcation: (a) idealised case of perfect symmetry; (b) disconnection due to infinitesimal imperfection; (c) massive disconnection due to realistic end conditions in the Taylor–Couette problem.

differential equations (see Guckenheimer & Holmes [25]). The outstanding challenge is to see whether these concepts can be applied to systems such as fluid mechanical problems, which are governed by partial differential equations.

In the case of Taylor–Couette flow, there are three important experimental parameters (see Fig. 1). The main one is the *Reynolds number Re*, which is defined as

$$Re = \frac{r_1 \omega d}{\nu}.$$

The term r_1 is the radius of the inner cylinder, ω is the angular rotation speed of that cylinder, d is the width of the gap between the cylinders and ν is the kinematic viscosity of the fluid. The Reynolds number is a measure of the dynamical state of the flow, with $Re = 0$ corresponding to the fluid being at rest. There are also two geometrical parameters. The *aspect ratio* Γ is defined as

$$\Gamma = \frac{h}{d},$$

where h is the distance between the two horizontal end plates. The *radius ratio* η is defined as

$$\eta = \frac{r_1}{r_2},$$

where r_2 is the radius of the stationary outer cylinder.

In practice, the radius ratio is a parameter that remains constant for any particular experimental configuration. Qualitative changes in the state of the flow are usually brought about through variation of the Reynolds number at fixed aspect ratio. Once a particular bifurcation has been located in this way, the measurement is repeated at different values of the aspect ratio in order to establish the path of that particular bifurcation in the two-dimensional parameter space given by (Re, Γ). Points at which two such paths intersect are codimension-2 points as discussed above. If one could also vary the radius ratio systematically, then these points would be seen to form loci in the three-dimensional parameter space of (Re, Γ, η). As will be seen in later sections, codimension-2 organising centres have an important role to play in controlling the chaotic behaviour that is found in Taylor–Couette flow.

3   Special Features of the Taylor–Couette Problem

There are several features of Taylor–Couette flow that merit further comment before proceeding to discuss the relevance of this work to turbulence or 'real world' flows. It must first be recognised that the observed sequence of well-defined instabilities is rather rare in fluid mechanics. Another case

where similar processes occur is in the problem of thermal convection between horizontal plates, known as *Rayleigh–Bénard flow*, but, more typically,
fluid flows become disordered after at most one instability. In addition, the
cellular structure formed at low rotation rates of the inner cylinder remains,
on average, to very high Reynolds numbers. Thus disordered motion is confined within a reasonably well-defined spatial structure. Much of this rich
structure can be attributed to the geometrical symmetries of the boundary
conditions, a topic which was first commented on by Terada & Hattori [56]
almost seventy years ago.

Taylor–Couette flow is an example of a 'closed flow', in which the effects of
instabilities are felt at all stations around the annular gap so that resonance-
type phenomena can be expected. Here we use the term 'closed flow' to
underline the fact that disturbances are fed back on to each other, reinforcing
some particular types of disturbances. This is in contrast to 'open flows',
such as flow in a pipe, where disturbances are carried off downstream and
may amplify or diminish as they are convected along. In addition to this,
the Taylor geometry of a circular cylinder rotating inside another has $SO(2)$
symmetry group such that only integral numbers of travelling waves are found
at the first temporal instability.

Another feature of Taylor–Couette flows is that the multiplicity of the solution set can be very high i.e. many different flows can coexist on the same
boundary conditions at the same Reynolds number. This was first demonstrated by Coles [13] in the time-dependent regime and by Burkhalter &
Koschmieder [8] and Benjamin & Mullin [5] for steady flows. Thus if we wish
to isolate individual bifurcation sequences to probe for explicit codimension-2
points then a good practical step is to restrict the physical size of the problem and thus reduce the rich multiplicity. This approach was first brought
to the fore in the pioneering work of T.B. Benjamin, who provided a theoretical framework on which such experimental observations could be based
(Benjamin [2]).

It might be thought that limiting the number of spatial modes in this
way will force low-dimensional behaviour and so will produce special types of
dynamical motion which are of limited scientific impact. However, this argument is misleading since the size of the system can be increased continuously
so that the chaos observed in small-scale systems is also found in extended
systems if sufficient care is taken in the observations, as shown by Pfister [43].
Other effects may also be found in a large system but great care is required
to distinguish them from dynamics caused by the practical difficulties associated with the very high multiplicity of the solution set. In any case, it has
been shown by Mullin [36] that both high and low-dimensional dynamics can
coexist in a small enclosed flow so that the infinite-dimensional aspect of the
Navier–Stokes equations is retained.

Finally, the appearance of steady cellular motion in the Taylor problem is a
dramatic event when viewed with an appropriate flow visualisation technique.

In one sense this creates difficulties for the subject since this striking process can be produced with very modest apparatus which need not be constructed too carefully. It is thus an excellent lecture demonstration, for example, but taking it from this forum to produce a basis for high-precision scientific study requires meticulous care in the control of the experiment. This is particularly true if one wishes to attempt an absolute comparison between numerical calculations of the Navier–Stokes equations on physical boundary conditions. This is an important issue, for if a computational fluid dynamics code is to have any value then it ought to be possible to calculate these examples of uncomplicated nonlinear flows.

4 Bifurcations in Taylor–Couette Flow

4.1 Appearance of steady cellular flows

Probably the greatest advance to date in hydrodynamic stability theory was made by Taylor [55], who carried out a theoretical and complimentary experimental study of the viscous form of the Rayleigh stability criterion. He showed that by rotating two concentric cylinders in an appropriate fashion, the circular Couette flow produced between them would become unstable to what we now call *Taylor cells* above some critical threshold. Even more remarkably, he also provided analytical estimates of the critical Reynolds number from a treatment of the equations of motion, and these were in very close agreement with the experimental observations.

The cylinders in Taylor's analytical model were assumed to be infinitely long. In this way, rotary Couette flow is a solution to the problem for all Reynolds numbers but it exchanges stability with a periodic cellular flow at a critical value of the parameter. The wavelength of the cells is close to the gap width between the cylinders, which is in agreement with experimental observations. Taylor also chose to make the radius ratio of the cylinders close to one so that he could obtain analytical solutions to the equations of motion. Nowadays, the nonlinear regime of the wide-gap problem has been extensively studied numerically and a good source of tabulated data can be found in DiPrima & Swinney [18].

The results of the analysis carried out by Taylor can be cast in modern mathematical terms in the form of a *pitchfork bifurcation*, as shown schematically in Fig. 3(a). In this model, the trivial branch corresponds to Couette flow and the supercritical branches of the pitchfork are then cellular flows. It is interesting to see how well this model fits experimental observations. For example, if we consider a system with an aspect ratio (length of cylinders divided by gap width) of forty then we would expect to see forty cells arise abruptly at a critical rotation rate of the inner cylinder. Thus if we take some measure of the flow at the mid-length of the cylinders we obtain the result

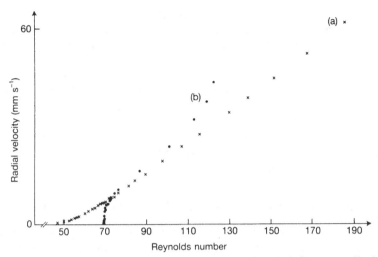

Figure 4: Experimental results showing onset of radial flow in a Taylor–Couette system with increasing Reynolds number half-way between the ends (circles) and close to one end (crosses).

shown in Fig. 4, where the measure we have used to distinguish the flows is the radial velocity component. It can be seen that there is an apparent abrupt transition to Taylor vortex flow which is in close agreement with the theoretical estimate. Also shown in Fig. 4 is a measure of the same property but now taken near one of the ends of the system. Now we see that the growth of the secondary flow is a smooth process with no obvious critical point. This reflects the fact that there is always some three-dimensional motion near the rigid end boundaries in an experiment and there is thus an element of choice in deciding when the whole flow field has undergone the critical event contained in the model. Thus the physical system cannot support the mathematical abstraction of rotary Couette flow along its entire length.

A popular modern treatment of this problem uses the so-called Ginsburg–Landau amplitude equation as a model. In this way both the temporal and spatial evolution of the bifurcating flow can be taken into account (see Cross & Hohenberg [14] for a comprehensive review of such ideas). This approach was used by Pfister & Rheberg [45] to account for the spatial inhomogeneity in their experimental observation of the development of an eighteen vortex flow as the Reynolds number was increased through the Taylor critical value. They obtained excellent agreement between theory and experiment, suggesting that the finite length effects of the smoothly evolving flow field are adequately modelled in this way.

The most obvious treatment of the ends is to model them by introducing an imperfection into the model such as in the Ginsburg–Landau approach discussed above. If the ends are considered as small effects on the whole flow field, then one might expect this representation to become more accurate as the length of the cylinders is increased. Thus the bifurcation will become disconnected as shown in Fig. 3(b). It can be seen that one of the solution branches evolves smoothly with increase in *Re* while the other is terminated at a saddle node so that it can only be reached with a discontinuous change in this parameter. This type of model seems to be in accord with the experimental results for the onset of cells discussed above. However, we have not yet considered experimental evidence for the disconnected branch of the pitchfork bifurcation, which is clearly an important part of the solution structure. This will be addressed in Section 5 below.

4.2   Hopf bifurcation to travelling waves

The first appearance of time-dependent motion in the Taylor–Couette system with increase of the Reynolds number is typically a travelling wave on top of the basic cellular structure. Flows of this type were first investigated in detail by Coles [13], who carried out an extensive experimental study. Davey, DiPrima & Stuart [17] applied weakly-nonlinear stability theory to the case of infinitely long cylinders to show that steady cellular flow becomes unstable to a travelling wave by means of a *Hopf bifurcation*. In this type of bifurcation, a stable fixed point exchanges stability with a limit cycle which is born at the bifurcation point. Two crucial properties that hold close to a Hopf bifurcation are that the amplitude of the oscillation grows as $|\mu - \mu_{crit}|^{\frac{1}{2}}$, where μ is the bifurcation parameter, and that the frequency of oscillation is independent of μ. Both properties have been found to hold experimentally (see Pfister & Gerdts [44], for example). Unlike the pitchfork bifurcation, the Hopf bifurcation is structurally stable to perturbations, and thus it has the same form in the infinite-cylinder model and the finite-cylinder experimental problem. This appears to be true despite the fact that no two cylinders are perfectly circular and every bearing has a small amount of clearance. Thus it is almost certain that all experiments contain a parasitic component at the frequency of rotation which could be thought of as a 'dynamic imperfection'. Indeed in narrow gap experiments such as those by Fenstermacher, Swinney & Gollub [19], there is a noticeable component at the driving frequency but this has no detectable effect on the results.

The exact details of the onset of travelling waves are found to be highly sensitive to the shape of the steady Taylor vortices at the point of instability as first pointed out by Mullin & Benjamin [38]. As has been mentioned already, the cells in the primary mode typically have square cross-section and hence the number of cells is very close to the aspect ratio of the annular gap. However, it is possible to stretch or compress the Taylor vortices by making careful adjustments to the aspect ratio once the primary mode is formed. Such a procedure was carried out experimentally by Mullin [34] for the case of a Taylor vortex flow comprising 26 cells. The results showed that the critical Reynolds number for the onset of time-dependence depends strongly on the aspect ratio, and that there is a change of wave number of the travelling wave as the aspect ratio is varied. Furthermore, Mullin showed that the qualitative form of the locus of Hopf bifurcations in the (Re, Γ) parameter space is itself highly dependent on the third variable experimental parameter, namely the radius ratio η. The role played by the radius ratio in determining the critical Reynolds number for the onset of travelling waves has been investigated numerically by Jones [26].

A significant development in the understanding of the Taylor–Couette problem came about through the work of Mullin, Cliffe & Pfister [39]. Using both experimental and computational techniques, they showed that for the specific case of 4-cell Taylor vortex flow it is possible to trace the origin of a Hopf bifurcation to the interaction between two steady-state bifurcations that exist in the problem. The temporal nature of this flow was novel in that it corresponded to an *axisymmetric* wave so that it was in phase around the cylindrical gap. A key ingredient of the bifurcation sequence were unstable solutions which were observable using numerical bifurcation techniques but which were inaccessible experimentally. By considering the respective stabilities of all the known solutions in a restricted region of parameter space, Mullin, Cliffe and Pfister demonstrated that the appearance of a Hopf bifurcation was required for completeness. Such a Hopf bifurcation was indeed found experimentally, and at critical parameter values that were in good agreement with those calculated numerically. These results provided the first evidence for codimension-2 behaviour in the Taylor–Couette problem and gave the indication that there were perhaps other types of multiple bifurcation points possible.

At first sight, the most obvious candidate arises from the 'perturbed' pitchfork bifurcation underlying the onset of cellular flows and the Hopf bifurcation that gives rise to travelling waves. One might imagine that these two bifurcations could interact as the second parameter, the aspect ratio, is varied. There would then be a codimension-2 bifurcation which would act as an organising centre for the dynamical chaos that exists in the problem. However, it will be shown below that such an argument is seriously flawed since the presence of the solid ends that bound the flow domain do not simply soften the steady bifurcation but destroy it.

5 Anomalous Modes

It seems remarkable now that the effects of realistic end conditions on Taylor–Couette flow were not considered until the work of Benjamin in the 1970's. While carrying out experiments to verify his theories of stability exchange in hydrodynamics, Benjamin [3] observed a type of Taylor vortex flow that had never previously been reported. The novelty was in the cells immediately adjacent to the stationary solid ends that defined the vertical extent of the flow domain. Conventional wisdom said that these cells should always circulate so that the radial flow next to the ends is directed towards the rotating inner cylinder. The physical argument that was used suggested that the diminishing of centrifugal effects towards the non-rotating ends would mean that an Ekman layer would form on each end wall such that there would be inward flowing fluid along them. However, Benjamin was able to create flows in which either one or both end cells circulated in the 'wrong' direction. Such flows were only stable for sufficiently high cylinder speeds, and collapsed back to one of the standard flows when the speed was reduced below a critical value. Taylor vortex flows of this type are now known collectively as *anomalous modes*. A full review of the role played by anomalous modes in the Taylor–Couette problem has been given by Cliffe, Kobine & Mullin [12].

The *odd-cell* states appear in pairs on symmetric boundary conditions and, with the exception of the single cell, are all disconnected solutions. They all exist at Reynolds numbers far in excess of those of interest here and so will not be discussed further. It was shown subsequently by Benjamin & Mullin [4] that the *even-cell* anomalous modes are in fact the missing branch of the pitchfork bifurcation described in Section 4. Using a model first proposed by Schaeffer [49], Benjamin and Mullin showed that the perfect pitchfork as shown in Fig. 3(a) corresponds to the case of a system with finite length which matches Couette flow at the ends. When an infinitesimal amount of 'friction' is included, the pitchfork bifurcation becomes disconnected as shown in Fig. 3(b). The connected branch corresponds to cellular flow with an inwardly-directed radial velocity component at the ends. The disconnected branch is the corresponding anomalous mode. Thus we have the situation that is found experimentally, namely that quasistatic increase of Re leads to the continuous onset of cellular flow, and that there is another disconnected flow which collapses catastrophically below a minimum critical value of Re.

However, it must be remembered that so far we have only included an infinitesimal amount of 'friction' at the ends. In a physically realistic system, there is a finite amount of 'friction' and this leads to a significant further change to the underlying pitchfork bifurcation. It has been found both experimentally and computationally that the disconnection of the pitchfork in real systems is extremely large. The splitting is relatively insensitive to cell number or aspect ratio but depends sensitively on the radius ratio η of the cylinders such that the disconnection is greater for narrow gaps. Typical values of the critical Reynolds numbers for the onset of cells and for the collapse of anomalous modes are $Re = 70$ and 180 respectively for $\eta = 0.5$, rising to $Re = 108$ and 715 respectively for $\eta = 0.85$. The effect of the ends is therefore not simply a perturbation of the periodic model, but represents a fundamentally different problem which must be studied in its own right. A more representative bifurcation diagram is therefore of the type shown in Fig.

3(c). There is no point in making experimental systems large to 'avoid end effects'. In fact the high multiplicity in the solution set that exists for long cylinders could mask more fundamental phenomena.

The reason that the disconnection is so large is that the pertinent symmetry of the periodic model is that of translational invariance. The physical boundary conditions are far removed from this mathematical abstraction as first discussed by Benjamin [2]. It is interesting to note that in the Rayleigh–Bénard problem the disconnection does appear to be accurately modelled by a disconnected pitchfork. For example, Wartenkin, Haucke, Lucas & Wheatley [58] showed that even in a small scale experiment with only two convection rolls the disconnection was not very large. The important symmetry in the convection problem arises due to the Boussinesq approximation as discussed by Benjamin [2]. Thus there is a fundamental difference between Rayleigh–Bénard and Taylor–Couette flows.

It is a remarkable fact that the periodic model, as first formulated by G.I. Taylor, gives an accurate estimate for the onset of cells which is sharp when experiments are carried out with long cylinders. The quadratic form of the primary branch constructed from a measure taken at the midplane of a long system therefore suggests that the effect of the imperfection is small. However, the other half of the 'bifurcation' is far away so that this cannot be considered as the simple softening of a pitchfork bifurcation caused by the small imperfection of distant solid boundaries.

One important consequence of the above result is that the argument for codimension-2 points presented at the end of Section 4 is flawed because the underlying pitchfork is effectively destroyed by the imposition of realistic boundary conditions at the ends. Certainly, it is still possible for a Hopf bifurcation to occur simultaneously with the saddle-node bifurcation that marks the lower stability limit of the disconnected branch. However, it would only be present at values of Re far in excess of those of interest here. Thus, as far as the primary solution branch is concerned, there appears to be no steady-state bifurcation with which the Hopf bifurcation can interact.

6 Symmetry Breaking

A solution to the above paradox is provided by considering the reflectional symmetry of the flow domain about the horizontal mid-plane. This symmetry is referred to mathematically as a Z_2 symmetry group. The cellular flows that appear with quasistatic increase of the Reynolds number from zero are always symmetric about the mid-plane. However, it is found both experimentally and computationally that these initially symmetric flows typically become asymmetric with a further increase of Re, *before* the occurrence of the Hopf bifurcation. Above a certain critical Re, cells in one half of the domain begin to grow at the expense of cells in the other half.

It is found that this process may be represented by a pitchfork or, as it is also known, *symmetry-breaking* bifurcation. Below the critical Re, the symmetric cellular flow solution is stable and can therefore be observed experimentally. Above the critical Re, it becomes unstable and stability is taken up by two asymmetric solution branches. These correspond to cellular flows that have broken the Z_2 symmetry of the flow domain. They occur in pairs because any asymmetric flow can be reflected about the midplane to give a second flow with the opposite sense of asymmetry.

The symmetric pitchfork bifurcation is a mathematical idealisation. In reality there will always be small imperfections which break the perfect Z_2 symmetry of the domain. In terms of the underlying bifurcation, these imperfections result in the disconnection of one of the asymmetric solution branches, as has been seen previously. However, in the case of the symmetry-breaking bifurcation for cellular flow, this disconnection is very small since the imperfections are themselves only small perturbations of the ideal case. The pitchfork bifurcation essentially remains complete, albeit in a disconnected state, as opposed to the case for anomalous modes where it is effectively destroyed. Thus it is extremely important to recognise the relevant physical symmetries of a problem and to treat those associated with mathematical abstractions with caution.

7 Low-Dimensional Chaos

With the discovery of symmetry-breaking bifurcations on the primary solution branch, we finally arrive at a situation where steady-state and Hopf bifurcations can interact in the parameter space formed by aspect ratio and Reynolds number to give organising centres for chaotic dynamics. The fact that the steady-state bifurcation is not from a trivial flow makes theoretical studies of such phenomena extremely complicated. Nevertheless, there are techniques in bifurcation theory, such as normal-form analysis (Carr [9]), that in principle can allow progress to be made in spite of such difficulties.

We will give two examples of occurrences of chaotic dynamics in a particular realisation of the Taylor–Couette problem where the presence of an organising centre in parameter space appears to be the crucial factor. The boundary conditions in this case are slightly different to those that are normally applied. It is still the case that there is a stationary outer cylinder and a rotating concentric inner cylinder. However, the end plates that bound the domain this time rotate with the inner cylinder instead of being stationary.

7.1 Interaction of saddle-node and Hopf bifurcations

The first example comes from a study carried out by Mullin, Tavener & Cliffe [41] into the mechanism for production of quasiperiodicity and irregularity in a flow that consists of four Taylor cells. Numerical bifurcation techniques were applied to discretised versions of the Navier–Stokes equations with the appropriate boundary conditions. These calculations showed that the initially symmetric 4-cell flow undergoes a symmetry-breaking bifurcation as the Reynolds number is increased. It is a more complicated bifurcation than the one discussed previously, since it is *subcritical* at the point of bifurcation. The stable asymmetric branches terminate in saddle-node bifurcations at a lower value of Re than that associated with the symmetry-breaking bifurcation.

If the system is initialised such that the flow is on the steady asymmetric state just above the saddle-node point then the computations show that an increase of Re gives rise to a Hopf bifurcation. A travelling wave with azimuthal wavenumber $m = 1$, which is sometimes referred to as a *tilt wave*, is created at this bifurcation point. The above bifurcation sequence is shown schematically in Fig. 5. It is found that as the aspect ratio is increased from $\Gamma \approx 3.7$, the critical Reynolds numbers for the saddle-node and Hopf bifurcations approach each other, becoming coincident at $Re = 106.2$ for $\Gamma = 4.34$. This is shown in Fig. 6, where the numerically-determined loci of the saddle-node (AB) and Hopf bifurcations (BC) are plotted. Point B in Fig. 6 therefore represents the location of a codimension-2 bifurcation in this particular problem, involving the simultaneous occurrence of two structurally-stable codimension-1 bifurcations.

Also plotted in Fig. 6 are experimentally determined points at which the flow becomes first quasiperiodic and then irregular. For a fixed value of the aspect ratio, increasing the Reynolds number results in both the subcritical pitchfork bifurcation and the Hopf bifurcation being encountered (points corresponding to these bifurcations were omitted from the plot for the sake of clarity). Increasing Re beyond the primary Hopf bifurcation for $\Gamma > \Gamma_E$ (where Γ_E is the aspect ratio associated with point E) leads to a regime of quasiperiodicity, where the initial singly-periodic oscillation develops a periodic modulation of its amplitude as a result of a secondary Hopf bifurcation. The locus of these secondary Hopf bifurcations is plotted as the line DE in Fig. 6. At slightly larger values of aspect ratio ($\Gamma > \Gamma_H$), the quasiperiodic flow is found to become irregular with a further increase of Re. The line DH consists of points at which this transition to irregularity was found to occur. It can be seen that the loci for the onset of modulation and irregularity bend round through DEF and DHG respectively. Thus the irregular regime can be entered by either increasing or decreasing Re.

7.2 Šil'nikov dynamics

The second example of chaotic behaviour comes from the same experimental system that was described in Section 7.1. Once again, it is an asymmetric 4-cell flow that is involved, with the underlying bifurcation structure being that

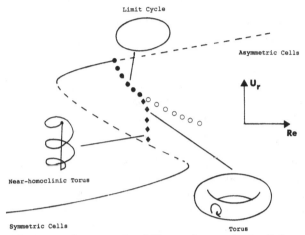

Figure 5: Bifurcation diagram of stability exchanges in 4-cell flow with variation of Reynolds number: stable limit cycle (black circles); unstable limit cycle (white circles); stable torus (black diamonds). Torus becomes homoclinic to unstable (dashed) asymmetric solution branch.

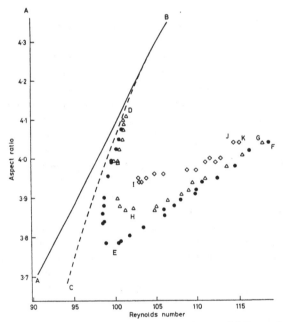

Figure 6: Parameter-space diagram showing lines of bifurcations in 4-cell flow: AB (solid), saddle node; BC (dashed), Hopf; DEF (black circles), secondary Hopf; DHG (white triangles), transition to irregularity; IJK (white diamonds), isola of regularly modulated flow.

uncovered experimentally and numerically by Mullin, Tavener & Cliffe [41]. In the course of investigating further the role played by the same codimension-2 organising centre as discussed in Section 7.1, Mullin & Price [40] uncovered a particular type of dynamical behaviour that has strong connections with an established low-dimensional model system.

The first example of the time series that were recorded experimentally by Mullin & Price [40] is shown in Fig. 7(a). The flow is in a dynamical state that consists of two different frequencies. The higher-frequency component corresponds to a travelling tilt wave of wavenumber $m = 1$, which breaks the continuous $SO(2)$ symmetry of the annular domain. The lower-frequency component is a wave of wavenumber $m = 0$ which is in phase at all points around the domain. The resulting state is one in which the higher-frequency component appears in discrete packets separated by relatively longer periods of approximately quiescent behaviour. It can be seen from the time series that the time between each wave packet is essentially constant, indicating that this is quasiperiodic motion.

The information contained in a one-dimensional time series can be portrayed in a multi-dimensional geometrical format by using the technique of *phase portrait reconstruction*. The details of this modern signal processing technique are beyond the scope of this article, and the interested reader is directed to the description given by Broomhead & Jones [7]. Here it is sufficient to know that the phase-space representation is achieved by extracting vectors from the time series in the form of short sequences of consecutive records. These vectors are then taken as the coordinates of points that sequentially form trajectories along which the system evolves with time. The overall collection of these trajectories forms an *attractor* in phase space that characterises the dynamics of the system.

The result of such a procedure applied to the time series in Fig. 7(a) is the reconstructed phase portrait shown in Fig. 7(b), which has the form of a torus. The best way to understand this diagram is to imagine a small marked particle that is constrained to move on the ribbon-like bands. As time evolves, so this marker moves along the prescribed trajectories. The quiescent phases of the experimental time series correspond to the marker moving along the central core of the phase portrait. It does this relatively slowly, and then spends a considerable time wandering erratically near one end. This erratic motion is almost certainly dues to measurement noise rather than dynamical noise from the fluid since the trajectory leaves this region rapidly at repeatable intervals. Thus at a well-defined time, when the marker reaches the end of the core, it is sent out rapidly on spiraling trajectories that initially expand in a plane perpendicular to the direction of the core. This expansion quickly saturates, and trajectories then spiral along until they rejoin the core at the far end, and the whole process repeats. The quasiperiodic nature of the dynamics can be seen in the way that the spiral trajectories are banded together. The fact that successive passes of trajectories in the bands are only slightly different is

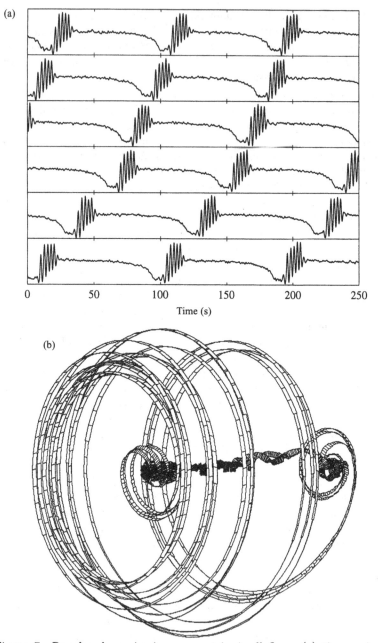

Figure 7: Regular dynamics in asymmetric 4-cell flow: (a) time series of radial velocity (strips read from left to right and from bottom to top); (b) reconstructed phase portrait.

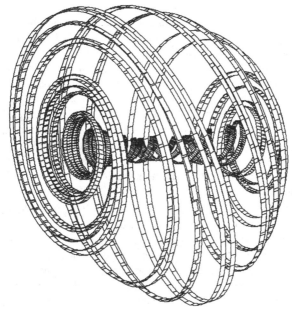

Figure 8: Regular phase portrait reconstructed from normal-form model proposed by Langford [29].

because the two frequencies which make up the torus are nearly commensurate with each other.

The virtue of investigating the experimental dynamics in this geometrical way is that it allows one to identify possible low-dimensional mechanisms that are responsible for controlling the observed behaviour. In this particular case, it would appear that a torus is approaching homoclinicity (see below) to an unstable fixed point at one end of the central core of trajectories. The high concentration of points at one end of the core indicates that the flow spends a relatively long time in this region of phase space so that the fixed point is *weakly* attracting along the nearly one-dimensional core. However it is strongly unstable with respect to spiraling motion in a plane orthogonal to the core. Near homoclinic motion on a torus of this type has been investigated in a particular normal form model by Langford [29]. We show a phase portrait obtained from the model in Fig. 8 where the qualitative features can be seen to be very similar to those observed in the experiment.

There is thus a correspondence between the experimental dynamics and those of certain ordinary differential equations that exhibit what is commonly referred to as the *Šil'nikov mechanism* for the onset of chaos. A detailed description of this low-dimensional mechanism can be found in the book by Guckenheimer & Holmes [25]. In essence, it involves motion in phase space that passes close to a saddle point which, by definition, has a stable and an unstable manifold associated with it. In the particular case studied by Šil'nikov [50], the saddle point is such that the motion close to the fixed point on one of the manifolds is a spiral on a sheet and the other is a line orientated normal to the sheet. Once a trajectory has been ejected along the unstable manifold, it is then reinjected via the stable manifold. A bifurcation parameter controls this reinjection processes, and it is found that *homoclinic* trajectories exist at particular values. A homoclinic trajectory is one that asymptotically approaches the same fixed point as time $t \to \pm\infty$. As such, motion on this type of trajectory has an infinite period. In the case of the Šil'nikov mechanism, it is found that there are various period-doubling cascades to chaotic motion associated with the existence of the homoclinic trajectories (see Mullin [37] for examples of this).

Returning to the experiments of Mullin & Price [40], it would appear that the dynamics of the flow are close to conditions of homoclinicity, since the period of motion around the attractor is considerably larger than the time scale set by the rotation of the inner cylinder. In fact, it is approximately ninety to one hundred times the rotation period of the inner cylinder. In addition, it was found that if one of the parameters was varied slightly from the value associated with the quasiperiodic behaviour illustrated in Fig. 7, then the flow dynamics were transformed into aperiodic motion. This is illustrated by the time series and reconstructed phase portrait shown in Fig. 9. The same characteristics of isolated bursts of oscillations separated by quiescent phases are retained, but now it can be seen from the time series (Fig. 9(a)) that the interval between packets is no longer regular. In phase space (Fig. 9(b)), the irregularity is seen in the spread of positions at which trajectories intersect the core, and in the filling in of the outer shell of the attractor. It should be noted that the length of the time series used to reconstruct the attractors shown in Figs. 7 and 9 are the same. This appearance of apparently chaotic behaviour with variation of a control parameter is consistent with the predictions of the Šil'nikov mechanism. The normal form studied by Langford is three-dimensional and cannot proceed to chaos in the way described here. Instead, the two frequencies of the torus must lock and period double and this route has been found by Mullin [35].

Thus, these observations represent a strong indication for a connection between the Navier–Stokes equations of fluid motion and properties of finite-dimensional dynamical systems of a nontrivial kind. However, there remain outstanding problems before any more definite statements can be made. The normal form used in the model has not been systematically derived from the Navier–Stokes equations using Lyapunov-Schmidt reduction techniques and so there are coefficients in the model that have not yet been identified. Some of them have been specified from a reduction of the discretised Navier–Stokes equations used in the numerical studies but they involve a combination of the two control parameters Re and Γ (Cliffe [11]). In addition, empirical information from the experimental data can be used to estimate the order of

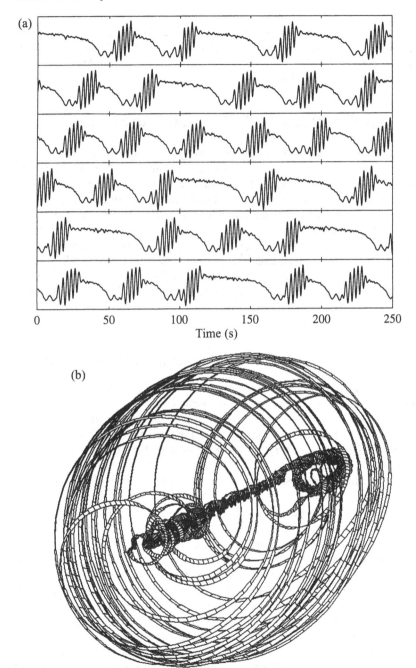

Figure 9: Chaotic dynamics in asymmetric 4-cell flow: (a) time series of radial velocity; (b) reconstructed phase portrait.

magnitude and the signs of some coefficients but their exact values remain unknown. In any case, the model is strictly only valid at the codimension-2 point and all of the observations discussed above were taken a finite distance away from the multiple bifurcation point.

8   Applications to Other Closed Flows

The above discussion has focused exclusively on bifurcations and dynamical processes in one particular fluid mechanical problem. However, for such ideas to have genuine significance in relation to understanding the Navier-Stokes equations, it is important to establish the validity of this approach for other types of flows. It has already been pointed out in Section 3 that the Taylor–Couette system as discussed so far contains a high degree of geometrical symmetry. In particular, there is continuous rotational symmetry in the azimuthal direction, which makes it a highly atypical example of a fluid flow. Also, Taylor–Couette flow is a closed flow, and as such it is to be expected that instabilities will occur at all locations simultaneously. This is unlike the case for open flows, where instabilities typically occur locally and are then convected with the flow.

8.1   Closed flows with discrete azimuthal symmetry

The role played by the continuous azimuthal symmetry of the Taylor–Couette problem in predisposing the system to display low-dimensional behaviour has been addressed recently in two experimental studies by the present authors. In both cases, the approach was to replace the stationary circular outer cylinder of the original geometry with one of non-circular cross-section, thereby breaking the continuous azimuthal symmetry. Pure travelling waves are no longer permitted under such conditions, and it is therefore of interest to see what forms of dynamical behaviour occur instead.

In the first study, by Kobine & Mullin [27], the new outer boundary was chosen to have a square cross-section. Mathematically, this results in the continuous $SO(2)$ symmetry group of the original problem being replaced with a discrete Z_4 group. Steady flows in such a system had already been investigated qualitatively by Snyder [52], who showed that the cellular flow had similar properties to those found in circular Taylor vortex flow. Dynamical phenomena were investigated by Kobine and Mullin in the regime of very small aspect ratios ($\Gamma \sim 1$), where the flow typically consists of just one cell. The onset of time-dependence with increasing Reynolds number was shown to have all the characteristics required of a Hopf bifurcation, as described in Section 4.2. Loci of several different codimension-1 bifurcations, including Hopf bifurcations, were mapped out experimentally in the control plane formed by aspect ratio and Reynolds number. Amongst those, a line of secondary

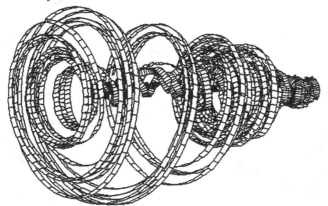

Figure 10: Phase-space reconstruction of dynamics in a Taylor–Couette flow with a square outer boundary.

Hopf bifurcations and a line of periodic folds (the time-dependent equivalent of a saddle-node bifurcation) were found to meet, forming a codimension-2 bifurcation. At parameter values close to this point, dynamical behaviour was found that gave rise to the phase portrait shown in Fig. 10 when reconstructed geometrically. The qualitative nature of such dynamics is clearly very similar to the behaviour discussed in Section 7.2 for the standard Taylor–Couette problem.

A second study, this time by Kobine, Mullin & Price [28], further reduced the degree of azimuthal symmetry by using an outer cylinder with a stadium-shaped cross-section. The cylinder consists of two parallel planar sides connected by two semicircular ends. The flow domain of the new system has discrete Z_2 symmetry in the azimuthal direction. Numerical investigations of the idealised two-dimensional velocity field showed that there is an approximately circular flow close to the rotating inner cylinder, with two large recirculations in the outer regions of the domain. Experimental investigations of the dynamical behaviour revealed that the onset of time-dependence with increasing Reynolds number was once again the result of a Hopf bifurcation to singly-periodic flow. A further increase of *Re* led to period-doubling bifurcations being encountered. An example of the first bifurcation from period-1 to period-2 behaviour can be seen in Fig. 11, where phase portraits that were reconstructed from the experimental dynamics are shown. The period-doubling cascade is one of the standard routes to chaos in low-dimensional dynamical systems (see Cvitanović [15], for example), and was first shown to occur in the standard Taylor–Couette problem by Pfister [43].

Critical values of *Re* for the Hopf bifurcation and the first period-doubling bifurcation were measured over a range of aspect ratio from $\Gamma = 3$ to 11 (the length scale was taken as the minimum gap between the inner and outer cylinders). The results showed the loci of the two bifurcations approaching

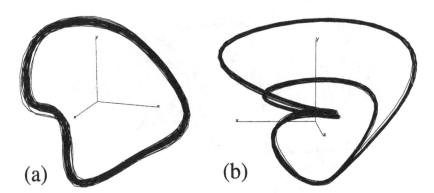

Figure 11: Example of period doubling in Taylor–Couette flow in a stadium-shaped domain: (a) period-1 limit cycle; (b) period-2 limit cycle.

each other with increasing Γ, and gave a strong indication of a codimension-2 bifurcation at $\Gamma \approx 7.5$. At aspect ratios greater than this value, the Hopf bifurcation to singly-periodic flow was still observed. However, on increasing *Re* further, it was found that there existed a sudden but reversible transition between the regular oscillatory behaviour and a type of motion that was highly irregular. Such a transition does not form part of the established framework of finite-dimensional dynamical systems theory. Nevertheless, it would appear that its existence is related to the probable coalescence of a Hopf bifurcation and a period-doubling bifurcation, and that the highly complicated phenomena that are observed in this problem are organised to an extent by this relatively simpler codimension-2 bifurcation.

The above theme was also extended in a different direction by an investigation of the effects of the connectedness of the fluid domain on the observed dynamics by Mullin [36]. Two small cylinders were positioned on the diameter of a large outer cylinder so that the flow was triply connected. The inner cylinders could be co-rotated or counter-rotated and cellular patterns of the Taylor vortex type were formed in some ranges of Reynolds number. One finding of this work was that the Ruelle-Takens route to chaos evolved as a particular solution branch was followed by quasistatically increasing the Reynolds number. In common with most fluid flows, there is multiplicity in the solution set so that other steady and time-dependent flows can exist on the same boundary conditions. In the present case, when one of the disconnected steady branches was followed it evolved into a high-dimensional state just above a Hopf bifurcation. The interesting point is that chaos and a high-dimensional irregular state coexisted as two of the attractors of the system. This observation challenges the argument that one should only ever expect to find low-dimensional chaos in small scale closed flows since there are only a few spatial modes present. The fluid is governed by the Navier–Stokes equations which are a set of partial differential equations and so restricting the spatial domain will not necessarily suppress high-dimensional dynamical motion.

8.2 Open flows

In contrast to the large number of reported occurrences of low-dimensional phenomena in closed flows, there have been very few observations made of such behaviour in open flows. However, there is a small but growing body of evidence to support the belief that the ideas that have been discussed throughout this article are also applicable to certain types of open flow. We will discuss three different examples.

An experimental study of the wake behind an oscillating cylinder was carried out by Olinger & Sreenivasan [42]. Without forcing, the wake developed the standard *Kármán vortex street*, involving the periodic shedding of eddies alternately from either side of the cylinder (see Tritton [57], for example). A local velocity measurement in such a wake gives a simple singly-periodic time series. However, when the cylinder was forced to move sinusoidally in a direction perpendicular to the free stream, a finite-dimensional dynamical response was produced. The natural frequency of the vortex street became coupled to the imposed frequency in a nonlinear manner. Olinger and Sreenivasan reported observations of locking between the wake and forcing frequencies, and demonstrated the existence of *Arnol'd tongues* in the parameter space of the system. Such behaviour can be related back to the fundamental properties of a simple finite-dimensional model known as the *sine circle map*. A review of these ideas applied to the Rayleigh–Bénard problem is given by Glazier & Libchaber [22].

Evidence for possible finite-dimensional behaviour in a boundary-layer flow has been given by Gaster [21]. The linear stability theory of two-dimensional boundary layers, in which individual modes become unstable and give rise to spatially and temporally periodic wave-like motion known as *Tollmien–Schlichting waves*, is well-known (see also Smith [51]). However, little is known about the subsequent nonlinear development of such waves at large amplitudes. Gaster's approach was to carry out experiments on flow over a flat plate in which there was a localised source of excitation. The dynamics of the flow were then recorded at various positions downstream of this source, corresponding to increasing Reynolds number with downstream distance. Excitation took the form of either sine-wave signals or noise, or combinations of the two. While the natural transition in the absence of excitation shows no obvious dynamical structure, the results obtained by Gaster showed that forcing leads to a significant subharmonic response and the development of broadband spectra at sufficiently large downstream distances. This is in accord with behaviour seen in coupled nonlinear systems of low dimension. Analysis of the experimental time series in terms of phase-space trajectories also showed evidence for low-dimensional behaviour. A review of the impact of dynamical systems theory on the problem of transition in boundary layer flows is given by Morkovin [33]. A specific model of intermittent phenomena in a boundary layer has been proposed by Aubry, Holmes, Lumley & Stone [1] and it contains some of the features of intermittency found by Price & Mullin [46] in the Taylor problem.

Arguably the most compelling evidence to date for the existence of finite-dimensional dynamics in an open-flow system comes from the experimental study carried out by Madden & Mullin [32] of flow in a planar channel which undergoes a symmetric sudden expansion. Upstream of the expansion, the flow, which is driven by a pressure gradient, is laminar and a good approximation to two-dimensional plane Poiseuille flow. As the flow emerges into the expanded region, it separates to form an ongoing jet and two confined recirculations. At low values of the Reynolds number, this configuration is symmetric about the midplane of the apparatus. However, with an increase of *Re*, there is a symmetry-breaking bifurcation to asymmetric flow, with one of the recirculations growing larger than the other. Under normal circumstances, the next qualitative change would be the development of time-dependent flow with a continuous spectral content centred on some mean frequency. However, Madden and Mullin showed that by modulating the overall flow rate sinusoidally, this high-dimensional behaviour is replaced by dynamics with definite low-dimensional characteristics. These include Hopf bifurcations, subharmonic response, hysteresis, homoclinicity and aperiodicity, all of which lie close to multiple bifurcation points in the control space.

As in the case of the periodically forced cylinder, it could be the case that adding an external drive changes the qualitative nature of the problem. However, even if this is the case, this example shows that structured chaos is relevant in an open flow. One practical consequence of this is that small amplitude forcing could be used to select low-dimensional dynamics of a preferred type in an open flow near transition. It should, however, be equally recognised that it is very unlikely that these approaches will be of any benefit in classical fluid problems of transition in a circular pipe. A recent study by Darbyshire & Mullin [16] found no evidence for either bifurcations or low-dimensional dynamical behaviour.

9   Conclusions

The ideas that have been discussed in this article have proved to be very successful in understanding complicated fluid flows and show that it may well be possible eventually to establish a formal connection between the Navier–Stokes equations and low-dimensional models. Certainly such ideas have been exploited most successfully in a number of cases to allow significant insights to be made into the mechanisms for the appearance of weakly turbulent fluid motion from initially laminar states. The Taylor–Couette problem in particular has proved an extremely useful system for developing many of the techniques that have been required, while at the same time yielding important information about the ability of a continuous fluid to display dynamics involving a discrete and small number of modes. The fact that irregular time-dependent behaviour can and does result from such apparently minimal circumstances is perhaps the most remarkable fact to emerge from this new conceptual framework.

The fundamental differences between open and closed flows are sufficiently important for us to draw attention yet again to the need for more progress towards establishing the existence of low-dimensional phenomena in open systems, since it is in such flows that the greatest challenge of understanding lies. In each of the three examples discussed in Section 8, the common element of externally-imposed forcing is required before dynamics with recognisable low-dimensional characteristics are observed. If the systems are allowed to undergo their natural transitions to disorder, then no such behaviour is encountered. It remains an unanswered question as to whether the imposition of forcing reveals an underlying mechanism of the flow, and ultimately of the Navier–Stokes equations themselves, or if it is creating an artificial problem that has no bearing on the physics of the unforced transitional process.

It is our belief that any 'ultimate' answers to the problem of fluid turbulence will only come to light when there has been a unification of the currently separate issues of dynamical and spatial processes. As such, the approach that treats the fluid as a dynamical system can never give the solution in itself, since it ignores the fact that the fluid is spatially extended. Likewise, spatial correlation techniques will always fall short of completeness because of the averaging out of the exact details of the dynamical evolution. A new paradigm is clearly required, but until then it would appear that the best way forward is to continue the process of systematically moving from the highly symmetric closed systems in which chaotic flows were first observed to gradually more generic flows, those in which symmetries are relaxed and spatial dependency is increased. At the same time, the theoretical models must be refined and developed to keep pace with the increasing complexity of the flows that they are intended to represent. An example of this can be seen in the recent emergence of *coupled map lattices*, in which simple one-dimensional iterative maps are coupled together to form spatially extended dynamical systems (Chaté & Manneville [10]). Although nothing more than model problems at present, it is by developing in such directions that real breakthroughs can be made.

Finally, it is interesting to note that in 1964, in his *Lectures on Physics*, Richard Feynman [20] wrote *"The next great era of awakening of human intellect may well produce a method of understanding the* qualitative *content of equations. Today we cannot. Today we cannot see that the water flow equations contain such things as [the turbulence] that one sees between rotating cylinders.*" In stating his views in this way, Feynman at once summed up the magnitude of the problem of solving the Navier–Stokes equations for anything but simple laminar flows, and at the same time provided a prophetic vision of how the impasse might be lifted. Thirty years later, with the growing utility of the ideas that arise from bifurcation theory and the theory of low-dimensional nonlinear systems, we may be seeing the first glimmers of Feynman's 'next great era'.

References

[1] Aubry, N., Holmes, P., Lumley, J.L. and Stone, E., The dynamics of coherent structures in the wall region of a turbulent boundary layer. *J. Fluid Mech.* **192**, 115–173, 1988.

[2] Benjamin, T.B., Bifurcation phenomena in steady flows of a viscous liquid. Part 1. Theory. *Proc. Roy. Soc. Lond. A* **359**, 1–26, 1978.

[3] Benjamin, T.B., Bifurcation phenomena in steady flows of a viscous liquid. Part 2. Experiments. *Proc. Roy. Soc. Lond. A* **359**, 27–43, 1978.

[4] Benjamin, T.B. and Mullin, T., Anomalous modes in the Taylor experiment. *Proc. Roy. Soc. Lond. A* **377**, 221–249, 1981.

[5] Benjamin, T.B. and Mullin, T., Notes on the multiplicity of flows in the Taylor–Couette experiment. *J. Fluid Mech.* **121**, 219–230, 1982.

[6] Brandstater, A. and Swinney, H.L., Strange attractors in weakly turbulent Couette–Taylor flow. *Phys. Rev. A* **35**, 2207–2220, 1987.

[7] Broomhead, D.S. and Jones, R., Time-series analysis. *Proc. Roy. Soc. Lond. A* **423**, 103–121, 1989.

[8] Burkhalter, J.E. and Koschmieder, E.L., Steady supercritical Taylor vortex flow. *J. Fluid Mech.* **58**, 547–560, 1973.

[9] Carr, J., *Applications of Centre Manifold Theory.* Applied Mathematical Sciences **35**, Springer, 1981.

[10] Chaté, H. and Manneville, P., Spatio-temporal intermittency in coupled map lattices. *Physica D* **32**, 409–422, 1988.

[11] Cliffe, K.A., Private Communication, 1995.

[12] Cliffe, K.A., Kobine, J.J. and Mullin, T., The role of anomalous modes in Taylor–Couette flow. *Proc. Roy. Soc. Lond. A* **439**, 341–357, 1992.

[13] Coles, D., Transition in circular Couette flow. *J. Fluid Mech.* **21**, 385–425, 1965.

[14] Cross, M.C. and Hohenberg, P.C., Pattern formation outside of equilibrium. *Rev. Mod. Phys.* **65**, 851–1112, 1993.

[15] Cvitanović, P., *Universality in Chaos*, Adam Hilger, 1989.

[16] Darbyshire, A.G. and Mullin, T., Transition to turbulence in constant-mass-flux pipe flow. *J. Fluid Mech.* **289**, 83–114, 1995.

[17] Davey, A., DiPrima, R.C. and Stuart, J.T., On the instability of Taylor vortices. *J. Fluid Mech.* **31**, 17–52, 1968.

[18] DiPrima, R.C. and Swinney, H.L., Instabilities and transition in flow between concentric rotating cylinders. In *Hydrodynamic Instabilities and the Transition to Turbulence* (eds. H.L. Swinney and J.P. Gollub). Topics in Applied Physics, Vol. 45, 139–180, Springer, 1981.

[19] Fenstermacher, P.R., Swinney, H.L. and Gollub, J.P., Dynamical instabilities and the transition to chaotic Taylor vortex flow. *J. Fluid Mech.* **94**, 103–28, 1979.

[20] Feynman, R.P., Leighton, R.B. and Sands, M., *The Feynman Lectures on Physics*, Ch. 41, Addison-Wesley, 1964.

[21] Gaster, M., The nonlinear phase of wave growth leading to chaos and breakdown to turbulence in a boundary-layer as an example of an open system. *Proc. Roy. Soc. Lond. A* **430**, 3–24, 1990.

[22] Glazier, J.A. and Libchaber, A., Quasiperiodicity and dynamical systems: an experimentalist's view. *IEEE Trans. Circ. Syst.* **35**, 790–809, 1988.

[23] Gollub, J.P. and Swinney, H.L., Onset of turbulence in a rotating fluid. *Phys. Rev. Lett.* **35**, 927–930, 1975.

[24] Golubitsky, M. and Schaeffer, D.G., *Singularities and Groups in Bifurcation Theory. Vol. 1.* Applied Mathematical Sciences **51**, Springer, 1985.

[25] Guckenheimer, J. and Holmes, P., *Nonlinear Oscillations, Dynamical Systems and Bifurcations of Vector Fields*, Second Edition, Applied Mathematical Sciences **42**, Springer, 1986.

[26] Jones, C.A., The transition to wavy Taylor vortices. *J. Fluid Mech.* **157**, 135–162, 1985.

[27] Kobine, J.J. and Mullin, T., Low-dimensional bifurcation phenomena in Taylor–Couette flow with discrete azimuthal symmetry. *J. Fluid Mech.* **275**, 379–405, 1994.

[28] Kobine, J.J., Mullin, T. and Price, T.J., The dynamics of driven rotating flow in stadium-shaped domains. *J. Fluid Mech.* **294**, 47–69, 1995.

[29] Langford, W.F., Numerical studies of torus bifurcations. *Int. Ser. Num. Math.* **70**, 285–293, 1984.

[30] Lorenz, E.N., Deterministic nonperiodic flow. *J. Atmos. Sci.* **20**, 130–141, 1963.

[31] Lumley, J.L., *Whither Turbulence? Turbulence at the Crossroads.* Lecture Notes in Physics, Vol. 357, Springer, 1990.

[32] Madden, F.N. and Mullin, T., An experimental observation of low-dimensional dynamics in an open-channel flow. *Phys. Fluids* **7**, 2364–2374, 1995.

[33] Morkovin, M.V., Recent insights into instability and transition to turbulence in open-flow systems. *NASA CR-181693*, 1988.

[34] Mullin, T., Onset of time-dependence in Taylor–Couette flow. *Phys. Rev. A* **31**, 1216–1218, 1985.

[35] Mullin, T., Finite-dimensional dynamics in Taylor–Couette flow. *IMA J. App. Math.* **46**, 109–120, 1991.

[36] Mullin, T., Disordered fluid motion in a small closed system. *Physica D* **62**, 192–201, 1993.

[37] Mullin, T., *The Nature of Chaos.* Oxford University Press, 1993.

[38] Mullin, T. and Benjamin, T.B., Transition to oscillatory motion in the Taylor experiment. *Nature* **288**, 567–569, 1980.

[39] Mullin, T., Cliffe, K.A. and Pfister, G., Unusual time-dependent phenomena in Taylor–Couette flow at moderately low Reynolds numbers. *Phys. Rev. Lett.* **58**, 2212–2215, 1987.

[40] Mullin, T. and Price, T.J., An experimental observation of chaos arising from the interaction of steady and time-dependent flows. *Nature* **340**, 294–296, 1989.

[41] Mullin, T., Tavener, S.J. and Cliffe, K.A., An experimental and numerical study of a codimension-2 bifurcation in a rotating annulus. *Europhys. Lett.* **8**, 251–256, 1989.

[42] Olinger, D.J. and Sreenivasan, K.R., Nonlinear dynamics of the wake of an oscillating cylinder. *Phys. Rev. Lett.* **60**, 797–800, 1988.

[43] Pfister, G., Deterministic chaos in rotational Taylor–Couette flow. In *Flow of Real Fluids* (eds. G.E.A. Meier and F. Obermeier), Lecture Notes in Physics, Vol. 235, 199–210, Springer, 1985.

[44] Pfister, G. and Gerdts, U., Dynamics of Taylor wavy vortex flow. *Phys. Lett. A* **83**, 23–25, 1981.

[45] Pfister, G. and Rheberg, I., Space-dependent order parameter in circular Couette flow transition. *Phys. Lett. A* **83**, 19–22, 1981.

[46] Price, T.J. and Mullin, T., An experimental observation of a new type of intermittency. *Physica D* **48**, 29–52, 1991.

[47] Ruelle, D., *Chaotic Evolution and Strange Attractors.* Cambridge University Press, 1989.

[48] Ruelle, D. and Takens, F., On the nature of turbulence. *Comm. Math. Phys.* **20**, 167–192, 1971.

[49] Schaeffer, D.G., Analysis of a model in the Taylor problem. *Math. Proc. Camb. Phil. Soc.* **87**, 307–337, 1980.

[50] Šil'nikov, L.P., A case of the existence of a denumerable set of periodic motions. *Sov. Math. Dokl.* **6**, 163–166, 1965.

[51] Smith, F.T., Weak and strong nonlinearity in boundary layer transition. This proceedings, ??-??.

[52] Snyder, H.A., Experiments on rotating flows between noncircular cylinders. *Phys. Fluids* **11**, 1606–1611, 1968.

[53] Swinney, H.L. and Gollub, J.P., *Hydrodynamic Instabilities and the Transition to Turbulence.* Topics in Applied Physics, Vol. 45, Springer, 1981.

[54] Tagg, R., The Couette–Taylor problem. *Nonlin. Sci. Today* **4**, 1–25, 1994.

[55] Taylor, G.I., Stability of a viscous liquid contained between two rotating cylinders. *Phil. Trans. Roy. Soc. Lond. A* **223**, 289–343, 1923.

[56] Terada, T. and Hattori, K., Some experiments on the motions of fluids. Part 4. Formation of vortices by rotating disc, sphere or cylinder. *Rep. Tokyo Univ. Aeronaut. Res. Inst.* **2**, 287–326, 1926.

[57] Tritton, D.J., *Physical Fluid Dynamics.* Oxford University Press, 1988.

[58] Wartenkin, P.A., Haucke, H.J., Lucas, P.G. and Wheatley, J.C., Stationary convection in dilute solutions of ^{3}He in superfluid ^{4}He. *Proc. Natl. Acad. Sci.* **77**, 6983–6987, 1980.

Weak and Strong Nonlinearity in Boundary Layer Transition

F. T. Smith

Department of Mathematics,
University College London,
London

1 Introduction

Our ultimate interest here is in instability and transition theory in boundary layers and related flows, including the effects of weak and strong nonlinearity as delineated later. We address this area via a gradual build-up through the following sections: Section 2, ordinary old boundary-layer theory; Section 3, nonlinear short-scale ideas; Section 4, unsteadiness and instability; Section 5, nonlinear unsteadiness/transition; Section 6, three-dimensional nonlinearity.

The main real applications and driving force over the years have been in aerodynamics, although the full range of real applications is in fact much wider. The theory to be described below applies in principle whenever the typical global Reynolds number Re is large, for flow near a solid surface. A helpful pictorial view of the type of instability and transition that are of concern here is presented in Figs. 104-106 of Van Dyke [46] which show photographs of the flow over a flat plate under various conditions in a near-uniform stream. At $Re = 20\,000$, with zero angle of incidence, the boundary-layer flow remains laminar, steady and approximately two-dimensional. When Re is increased to $100\,000$ however, small two-dimensional waves appear travelling downstream: these are the so-called Tollmien-Schlichting waves. Further, a slight change in the conditions, from zero to 1° angle of incidence, induces transition of the flow from the laminar state to a complex turbulent one, again at $Re = 100\,000$. The experimental progression in Fig. 106 in particular tends to tie in with a theoretical view based on small linear disturbances upstream, followed by weakly nonlinear and then strongly nonlinear processes downstream as transition to turbulent motion takes place. Along with that the flow changes from being two-dimensional upstream to three-dimensional downstream. The present article, at least through its contained list of references, touches later on some of these observations, given that Re is typically large.

The starting point is the Navier-Stokes equations for the fluid motion, given by

$$u_x + v_y = 0, \qquad (1.1a)$$

$$u_t + u u_x + v u_y = -p_x + Re^{-1} \left(u_{xx} + u_{yy} \right), \qquad (1.1b)$$

$$v_t + u v_x + v v_y = -p_y + Re^{-1} \left(v_{xx} + v_{yy} \right). \qquad (1.1c)$$

Here two-dimensional unsteady motion for an incompressible fluid of constant density ρ_D is assumed first, with the corresponding three-dimensional motion being considered towards the end. The equations are written in a nondimensional form based on a characteristic length scale ℓ_D and velocity scale u_D of the flow (see Fig. 1), the unknown velocity components u, v being in the Cartesian directions x, y and p being the unknown pressure based on $\rho_D u_D^2$. The time t is based on $\ell_D u_D^{-1}$. The Reynolds number Re is then $u_D \ell_D \nu_D^{-1}$ where ν_D is the kinematic viscosity of the fluid. The boundary conditions include

$$u = v = 0 \quad \text{at the body surface}, \qquad (1.2a)$$

$$u \to 1, \quad v \to 0 \quad \text{in the farfield}, \qquad (1.2b)$$

for the case of external flow past a given fixed body. There is much practical and theoretical interest in the solution properties for large Re, as hinted in the previous paragraph. One approach is to simply try to compute directly approximate solutions for (1.1a) - (1.2b) (so-called direct numerical simulations, e.g. Kleiser & Zang [21]) but these become extremely difficult at increased Re, especially in three dimensions. An alternative, which we describe here, is to construct a theory for $Re \gg 1$, to see if it may help the overall understanding of the system as well as suggesting more appropriate computational methods

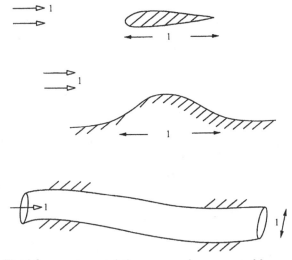

Figure 1: Fluid flow past an airfoil, over a surface-mounted hump, or through a pipe.

(e.g. Smith, Papageorgiou & Elliott [30], Bertolotti, Herbert & Spalart [2]). When Re is large, it is very tempting of course to just discard the Re^{-1} terms (viscous terms) in (1.1b,c) immediately. However, that is not usually correct as we shall see in the next section. Indeed, part of our aim in the present article is to derive the correct equations formally in a rational logical way rather than merely picking equations in an ad hoc fashion.

2 Ordinary Old Boundary Layer Theory

The classical theory for $Re \gg 1$ proceeds essentially in two steps which are described below, followed by some comments.

2.1 The majority of the flowfield

In most places, where x, y are typically of $\mathcal{O}(1)$, the flow solution is sought in the form

$$u = u_0 + Re^{-1/2}u_1 + \cdots, \tag{2.1a}$$

$$v = v_0 + Re^{-1/2}v_1 + \cdots, \tag{2.1b}$$

$$p = p_0 + Re^{-1/2}p_1 + \cdots, \tag{2.1c}$$

for steady motion with zero $\partial/\partial t$; unsteady motion is also covered below. Substitution into (1.1a-c) and balancing of the leading-order terms therefore yields the Euler equations

$$u_{0x} + v_{0y} = 0, \tag{2.2a}$$

$$u_0 u_{0x} + v_0 u_{0y} = -p_{0x}, \tag{2.2b}$$

$$u_0 v_{0x} + v_0 v_{0y} = -p_{0y}, \tag{2.2c}$$

for $(u_0, v_0, p_0)(x, y)$. In effect the viscous terms in (1.1b,c) are neglected as expected. That however drops the highest derivatives from the system, reducing its order to that of (2.2a-c) and so reducing the number of boundary conditions that can be satisfied. Instead of (1.2a) a condition of zero penetration, i.e. tangential flow, is usually guessed as being appropriate at the body surface ($y = f(x)$ say), so that

$$v_0 = u_0 f'(x) \quad \text{at} \quad y = f(x) \tag{2.3a}$$

replaces (1.2a) (see Fig. 2), while (1.2b) stays intact in the sense that we require

$$u_0 \to 1, \quad v_0 \to 0 \quad \text{in the farfield.} \tag{2.3b}$$

The Euler problem (2.2a-c), (2.3a,b) usually acts to determine $(u_0, v_0, p_0)(x, y)$ everywhere outside the body.

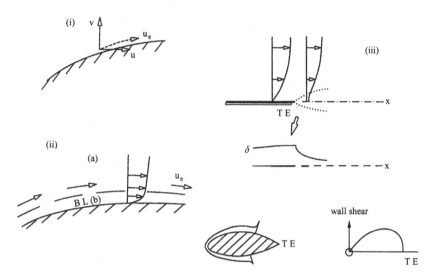

Figure 2: Flow near a solid surface: (i) inviscid, (ii) the $\mathcal{O}(\mathrm{Re}^{-1/2})$ boundary layer, (iii) failure near a trailing edge or at the onset of flow reversal.

Several points should be noticed at this stage. First, (2.2a-c) can often be reduced further, to the equations of potential flow. Second, the inviscid outer solution (u_0, v_0, p_0) above leaves a tangential slip velocity u_e near the body surface, given by

$$u_e \equiv \left\{1 + (f'(x))^2\right\}^{1/2} u_0 \qquad \text{at } y = f(x), \qquad (2.4)$$

which is nonzero in general. The boundary layer in Section 2.2 is needed to reconcile (2.4) with the zero-slip condition of (1.2a). Third, the surface pressure is also determined at leading order by the inviscid outer solution (p_0 at $y = f(x)$), prior to the inner solution in Section 2.2. Fourth, the outer solution depends on the local *slope* as in (2.3a). Fifth, the account (2.1a)-(2.4) is for steady flow, as a first go. Unsteady flows are broadly similar, adding u_{0t}, v_{0t} to (2.2b,c) for instance, unless the unsteadiness introduces its own extra scales to be accommodated.

2.2   The boundary layer

Near the body surface a so-called boundary layer is required. This is to restore the highest derivative or viscous term(s) in order for zero slip as in (1.2a) to hold at the surface rather than slip as in (2.4). The boundary layer is assumed to be thin (see Fig. 2).

The example of a flat plate aligned with the external stream is a basic one to start with. For then the outer solution is simply

$$(u_0, v_0, p_0) \equiv (1, 0, p_\infty), \tag{2.5}$$

where p_∞ is constant for steady motion. So here $u_e = 1$, with $f(x) = 0$ on the plate occupying $0 < x < 1$ say. The thin boundary layer therefore has the normal scaling

$$y = Re^{-1/2}\bar{y}, \tag{2.6}$$

with $\bar{y}$ of order unity. The reason or guess for the $\frac{1}{2}$ power in (2.6) is the balancing of uu_x (of order $1^2/1$) with the viscous term $Re^{-1}u_{yy}$ (of order $Re^{-1}.1/y^2$), in (1.1b). The boundary layer also has

$$u = \bar{u} + \cdots, \tag{2.7a}$$

$$v = Re^{-1/2}\bar{v} + \cdots, \tag{2.7b}$$

$$p = \bar{p} + \cdots, \tag{2.7c}$$

with $(\bar{u}, \bar{v}, \bar{p})$ being unknown functions of $x, \bar{y}$ and possibly t. Substituting into (1.1a-c) and balancing now yields, in turn,

$$\bar{u}_x + \bar{v}_{\bar{y}} = 0, \tag{2.8a}$$

$$\bar{u}_t + \bar{u}\bar{u}_x + \bar{v}\bar{u}_{\bar{y}} = -\bar{p}_x + \bar{u}_{\bar{y}\bar{y}}, \tag{2.8b}$$

$$0 = -\bar{p}_{\bar{y}}. \tag{2.8c}$$

Here the dominant viscous term $(\bar{u}_{\bar{y}\bar{y}})$ now appears at leading order in (2.8b), while (2.8c) shows the typical thin-layer result that the normal pressure gradient must be zero. The latter implies that $\bar{p} = \bar{p}(x, t)$ is independent of $\bar{y}$ and so matching with the outer expansion in (2.1c), with (2.5), requires $\bar{p} = p_\infty$ also. Hence $\bar{p}_x$ is also negligible in this example, leaving the x-momentum equation (2.8b) as

$$\bar{u}_t + \bar{u}\bar{u}_x + \bar{v}\bar{u}_{\bar{y}} = (\text{zero}) + \bar{u}_{\bar{y}\bar{y}}, \tag{2.8b'}$$

where our concern in this paragraph is with steady flow where $\partial/\partial t$ is zero. The appropriate boundary conditions on (2.8b') and the continuity equation (2.8a) are

$$\bar{u} = \bar{v} = 0 \quad \text{at} \quad \bar{y} = 0, \tag{2.8d}$$

$$\bar{u} \to 1 \quad \text{as} \quad \bar{y} \to \infty, \tag{2.8e}$$

from (1.2a) and from matching with (2.5) outside. These three conditions are in line with the order of (2.8a,b'). A starting profile at $x = 0+$ is also needed, this being $\bar{u} \equiv 1$ for all $\bar{y} > 0$. A solution of (2.8a,b',d,e) is then the Blasius similarity solution (Jones & Watson [20]) $\bar{u} = F'(\bar{y}/x^{1/2})$ where F satisfies a nonlinear ordinary differential equation subject to two-point boundary conditions covering (2.8d,e). Blasius' form is found to agree with experimental findings at moderate Re values.

For more general body shapes the (classical) idea is the same as in (2.6)-
(2.8e). In essence (2.8e) is replaced then by $\bar{u} \to u_e(x)$ in view of (2.4), while
(zero) in (2.8b') is replaced by $(-\bar{p}_x =)u_e u'_e$. Likewise for unsteady flows (2.8e)
becomes

$$\bar{u} \to u_e(x,t) \quad \text{as} \quad \bar{y} \to \infty, \tag{2.8f}$$

together with the replacement

$$(u_{et} + u_e u_{ex}) \quad \text{instead of (zero) in (2.8b').} \tag{2.8g}$$

The system (2.8a,b',d,f,g) for the boundary layer is parabolic in x generally. It
is written above for the upper surface of the body, x denoting distance along
the surface, and a similar account deals with the lower surface if necessary.

2.3   Comments

All looks well with the two steps described in Sections 2.1 and 2.2 at first
sight. However, the classical idea above fails in at least three ways (see Fig.
2). One is for the basic flat-plate example, where the solution is found to
produce quite strong discontinuities (e.g. in the slope of the displacement δ)
just beyond the plate's trailing edge (Goldstein [10], Messiter [22], Stewartson
[39]). Second, on more general bodies such as airfoils the solution usually fails
to reach the trailing edge anyway, as it encounters a singularity at a point of
zero wall shear on the body (Goldstein [11], Stewartson [41], Smith & Daniels
[28]). This is associated (loosely) with separation of the boundary layer from
the surface. Third, and perhaps of most concern here, unsteady disturbances
are found to be stable in general according to this method and so the classical
approach does not show/capture Tollmien-Schlichting waves for instance and
so is unsuitable for the study of transition (see also Section 1 and Smith [26]).

These failures, which mean that the idea described in Sections 2.1 and
2.2 does not often work over an entire body, point us towards an alternative
approach which is considered in the next section.

3   Nonlinear Short-Scale Ideas

We guess that a relatively short length scale (ℓ say) is needed locally in the
streamwise direction x, near a station x_0, to cope with unsteady waves, or
trailing edges, or separations, or whatever. The scale ℓ is assumed to exceed
the incoming boundary-layer thickness, so that $1 \gg \ell \gg Re^{-1/2}$, and ℓ is
determined by an ordering argument as follows (see also Fig. 3).

Since $\ell \ll 1$ local viscous forces are significant in a viscous sublayer thinner
than $\mathcal{O}(Re^{-1/2})$. Supposing the sublayer to have $y \sim \epsilon Re^{-1/2}$, with ϵ small
and unknown, the typical incoming u value there is of order ϵ in view of the
boundary layer in Section 2.2. This u is to be altered nonlinearly. Hence the

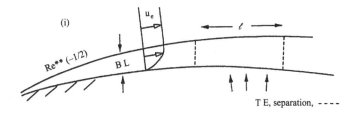

(i)

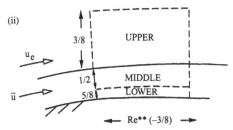

(ii)

Figure 3: Short-scale processes: (i) with unknown length scale $\ell \ll 1$, (ii) the triple-deck structure.

balance uu_x against $Re^{-1}u_{yy}$ in (1.1b) requires $\epsilon^2/\ell \sim Re^{-1}.\epsilon/(\epsilon^2 Re^{-1})$ (since $|x - x_0| \sim \ell$), giving

$$\ell \sim \epsilon^3, \tag{3.1}$$

while inclusion of p_x implies that $p \sim u^2$ or

$$\text{inner}-p \sim \epsilon^2. \tag{3.2}$$

Simultaneously the main boundary-layer thickness is altered by $\mathcal{O}(\epsilon Re^{-1/2})$ because of the sublayer thickness. So, outside the main boundary layer, a pressure is produced proportional to the slope, akin to (2.3a), i.e. of order $\epsilon Re^{-1/2}/\ell$, giving

$$\text{outer}-p \sim Re^{-1/2}\epsilon\ell^{-1}. \tag{3.3}$$

Consequently a new effect, namely *inner-outer interaction*, can occur when the pressures in (3.2) and (3.3) are equal. This interactive effect is the major difference from the old theory in Section 2 in which the order Section 2.1-then-Section 2.2 holds. Here, instead, they affect each other. From equating (3.2), (3.3) and from (3.1) we obtain the estimate $\epsilon = Re^{-1/8}$ and so

$$\ell = \mathcal{O}(Re^{-3/8}) \tag{3.4}$$

determines the new short length scale.

The orders (3.1)-(3.4) then form the basis for the expansions of the flow solution in the three regions or decks of the local triple-deck structure which is inferred from the argument above (see Fig. 3). We set $x - x_0 = \epsilon^3 X$.

In the *lower deck* or viscous sublayer, $y = \epsilon^5 Y$ and

$$u = \epsilon U + \cdots , \tag{3.5a}$$

$$v = \epsilon^3 V + \cdots , \tag{3.5b}$$

$$p = \epsilon^2 P + \cdots . \tag{3.5c}$$

So here (1.1a-c) yield to leading order the controlling equations of continuity and x-momentum,

$$U_X + V_Y = 0, \tag{3.6a}$$

$$U U_X + V U_Y = -P_X(X) + U_{YY}, \tag{3.6b}$$

respectively, whereas the y-momentum balance gives $\partial P/\partial Y = 0$ as in (2.8c), leaving $P = P(X)$ unknown. The typical boundary conditions on (3.6a,b) include

$$U = V = 0 \quad \text{at} \quad Y = 0, \tag{3.6c}$$

$$U \sim \lambda(Y + A(X)) \quad \text{as} \quad Y \to \infty, \tag{3.6d}$$

$$P(X) = \frac{u_e^2}{\pi} \int_{-\infty}^{\infty} \frac{A'(s)ds}{(X - s)}. \tag{3.6e}$$

Here (3.6c) stems directly from (1.2a) of course on a flat surface but is replaced by $U_Y = V = 0$ in the wake (say for $X > 0$) for a local trailing-edge flow. Conditions (3.6d,e) in contrast come from matching with the main- and upper-deck solutions below.

In the *main deck* spanning the majority of the original boundary layer, $y = \epsilon^4 \bar{y}$ with

$$u = \bar{u}_B(\bar{y}) + \epsilon \bar{u}_1 + \cdots , \tag{3.7a}$$

$$v = \epsilon^2 \bar{v}_1 + \cdots , \tag{3.7b}$$

$$p = \epsilon^2 P(X) + \cdots , \tag{3.7c}$$

where $\bar{u}_B(\bar{y})$ is the incoming boundary-layer velocity profile obtained as in Section 2.2 at $x = x_0 -$. From (1.1a-c) we therefore obtain the inviscid system

$$\bar{u}_{1X} + \bar{v}_{1\bar{y}} = 0, \tag{3.8a}$$

$$\bar{u}_B \bar{u}_{1X} + \bar{v}_1 \bar{u}_B'(\bar{y}) = 0, \tag{3.8b}$$

with (1.1c) confirming that the $\mathcal{O}(\epsilon^2)$ pressure term is independent of $\bar{y}$ and hence equal to $P(X)$ as in the lower deck. The solution here is simply

$$\bar{u}_1 = A(X)\bar{u}_B'(\bar{y}), \quad \bar{v}_1 = -A'(X)\bar{u}_B(\bar{y}), \tag{3.8c}$$

where the negative displacement function $A(X)$ is to be found and the positive constant $\bar{u}_B'(0) \equiv \lambda$. Matching with the lower-deck form then gives the condition (3.6d).

The *upper deck* outside the boundary layer has $y = \epsilon^3 y_*$ and

$$u = u_e + \epsilon^2 u_* + \cdots , \tag{3.9a}$$

$$v = \epsilon^2 v_* + \cdots , \tag{3.9b}$$

$$p = \epsilon^2 p_* + \cdots , \tag{3.9c}$$

as implied by the main-deck behaviour at large $\bar{y}$ where $\bar{u}_B \to u_e$, the constant local inviscid slip velocity (from Section 2.1) at $x = x_0\pm$. Substitution into (1.1a-c) now leads to Laplace's equation for the (inviscid outer) pressure disturbance,

$$\left(\frac{\partial^2}{\partial X^2} + \frac{\partial^2}{\partial y_*^2} \right) p_* = 0, \tag{3.10a}$$

subject to suitable boundedness in the farfield and

$$p_* \to P(X), \quad \frac{\partial p_*}{\partial y_*} \to u_e^2 A''(X) \quad \text{as} \quad y_* \to 0+, \tag{3.10b}$$

from matching with the main-deck solution. The solution of (3.10a,b) imposes the relation quoted earlier in (3.6e) between the unknown pressure and displacement.

We are left therefore with the task of solving the so-called interactive boundary-layer system (3.6a-e). Alterations of the surface conditions in (3.6c) allow the same system to deal with separations, various trailing-edge flows, flows over humps, dents, corners, steps, injection slots, and so on (see Fig. 4). Also, a normalisation allows u_e and λ to be set to one.

In preparation for Section 4 we observe that for hump flow for example, $U = V = 0$ at $Y = hG(X)$ instead of (3.6c), with the hump shape G being of $\mathcal{O}(1)$, bounded and smooth say, and h representing a height parameter. Linearised solutions for small h are then useful, giving

$$(U, V, P, A) = (Y, 0, 0, 0) + h(U_1, V_1, P_1, A_1) + \mathcal{O}(h^2), \tag{3.11}$$

so that (3.6a,b) become, at order h,

$$U_{1X} + V_{1Y} = 0, \tag{3.12a}$$

$$YU_{1X} + V_1 = -P_1'(X) + U_{1YY}. \tag{3.12b}$$

Differentiation of (3.12b) with respect to Y and use of (3.12a) therefore leaves the linear equation $Y\partial\tau_1/\partial X = \partial^2\tau_1/\partial Y^2$ for the shear perturbation $\tau_1 \equiv \partial U_1/\partial Y$. This equation becomes Airy's equation

$$\frac{\partial^2 \tau_1^F}{\partial Y^2} = i\sigma Y \tau_1^F \tag{3.13a}$$

after a Fourier transform (denoted by F) is applied in X, yielding the solution

$$\tau_1^F \propto Ai\left\{ (i\sigma)^{1/3} Y \right\}, \tag{3.13b}$$

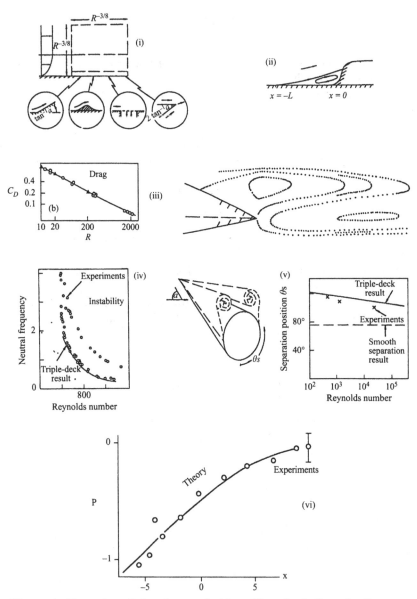

Figure 4: Examples of the short-scale ideas (i) and solutions for flow near steps (ii), trailing edges (iii), instability (iv), 2D and 3D separation (v,vi).

where σ is the transform variable. The rest of the working is as in Stewartson [42] and Smith [25]. At $\mathcal{O}(1)$ values of h, or for trailing edges or breakaway separating flows, careful computation is necessary in general (see Stewartson & Williams [40, 43], Smith [29], Messiter [23]).

The short-scale ideas applied in this section to local steady flows are extended next to unsteady motions.

4 Unsteadiness and Instability

Exactly the same arguments as in Section 3 hold for short-scale effects with unsteadiness present provided that we compare $|\partial/\partial t|$ with $|u\partial/\partial x|$, in view of the momentum equations (1.1b,c). The magnitude of the operator $u\partial/\partial x$ in Section 3 is at its smallest within the lower deck however, since there u is $\mathcal{O}(\epsilon)$ (in (3.5a)) as opposed to $\mathcal{O}(1)$ in the main and upper decks in (3.7a), (3.9a). So the suggested crucial time scale t is of order $(x - x_0)/\epsilon$, or

$$t = \epsilon^2 T, \qquad (4.1)$$

where $\epsilon \equiv Re^{-1/8}$ still. The temporal scaling (4.1) with T of order unity leads effectively to only one change in the system (3.6a-e), adding U_T to (3.6b) to produce

$$U_T + UU_X + VU_Y = -P_X(X,T) + U_{YY}. \qquad (4.2)$$

Now P and A are unknown functions of X and T, and U and V depend on X, Y and T. In addition an initial condition, say at $T = 0$, may need to be set. Otherwise, the system of concern here is (3.6a,c-e), (4.2), for two-dimensional unsteady motion.

Linearised properties are examined first now. The linearisation is as in (3.11), signifying a small perturbation of the original boundary-layer motion $U \equiv Y$. Hence, with U_{1T} added to (3.12b) to accommodate (4.2), the linear equation for τ_1 becomes

$$\tau_{1T} + Y\tau_{1X} = \tau_{1YY}. \qquad (4.3)$$

If wave-like dependence $\propto \exp(i(\alpha X - \Omega T)) \equiv \hat{E}$ is sought, for constant wavenumber α and frequency Ω, then (4.3) becomes Airy's equation again, c.f. (3.13a). Accordingly, with Ai again denoting the Airy function,

$$\hat{\tau}_1 = BAi(\xi), \quad \xi \equiv (i\alpha)^{1/3}(Y - \Omega/\alpha), \qquad (4.4a)$$

where $\tau_1 = \hat{\tau}_1(Y)\hat{E} + c.c.$ etc. (c.c. denotes the complex conjugate), and B is an unknown constant. Including, if necessary, forcing conditions due to a moving wavy wall for instance, $\hat{U}_1 + \hat{G} = \hat{V}_1 + i\Omega\hat{G} = 0$ at $Y = 0$, where $\hat{G}$ is a wall-shape constant, equation (3.12b) (plus U_{1T}) requires $\hat{\tau}_{1Y} = i\alpha\hat{P}_1$ at $Y = 0$. Hence from (4.4a), with ξ_0 denoting the value of ξ when Y is zero,

$$B(i\alpha)^{1/3}Ai'(\xi_0) = i\alpha\hat{P}_1. \qquad (4.4b)$$

On the other hand, integrating $\hat{\tau}_1$ to obtain $\hat{U}_1(Y)$, with $\hat{U}_1(0) = -\hat{G}$, and then imposing (3.6d), i.e. $\hat{U}_1(\infty) = \hat{A}_1$, gives

$$B(i\alpha)^{-1/3} \int_{\xi_0}^{\infty} Ai(q)dq - \hat{G} = \hat{A}_1, \qquad (4.4c)$$

while the interaction law (3.6e) becomes, with $\mathrm{Re}(\alpha)$ assumed positive,

$$\hat{P}_1 = \alpha\hat{A}_1; \qquad (4.4d)$$

the latter can be derived alternatively from (3.10a,b). It follows that (4.4b-d) provide three equations for B, $\hat{P}_1$ and $\hat{A}_1$. In particular the perturbed pressure solution is given by

$$\left\{(i\alpha)^{4/3}I - iAi'(\xi_0)\right\}\hat{P}_1 = \hat{G}i\alpha Ai'(\xi_0), \qquad (4.5)$$

where I is the integral appearing in (4.4c).

The solution (4.5) can be used in interpreting forced or receptivity problems. The chief point however is that waves are possible when the left-hand coefficient in (4.5) vanishes, giving resonance for nonzero $\hat{G}$. The waves' dispersion relation is

$$(i\alpha)^{2/3} Ai'(\xi_0) = i\alpha^2 \int_{\xi_0}^{\infty} Ai(q)dq, \qquad (4.6a)$$

where $\xi \equiv -i\Omega/(i\alpha)^{2/3}$. These waves are usually considered for real Ω, the input frequency imposed by a vibrating ribbon upstream for instance, in

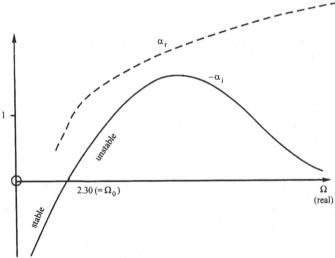

Figure 5: Wavenumber α_r, spatial growth rate $-\alpha_i$, versus frequency Ω, in scaled terms.

which case α, determined by (4.6a), is generally complex. The dependence of the real and imaginary parts (α_r, α_i) of α on Ω is sketched in Fig. 5. Neutral stability, corresponding to zero α_i, is attained for the particular value $\Omega = \Omega_0$, for which $\alpha = \alpha_0$, where

$$\Omega_0 \approx 2.30, \quad \alpha_0 \approx 1.00. \tag{4.6b}$$

The waves are stable for $\Omega < \Omega_0$ but unstable for $\Omega > \Omega_0$ (Smith [26]). They are the *Tollmien-Schlichting waves* mentioned in Section 1. More precisely they are the lower-branch such waves, for large *Re*.

For later use we observe that the instability persists at high Ω values but with

$$\alpha^2 \sim \Omega \quad \text{for } \Omega \gg 1, \tag{4.6c}$$

from (4.6a), then α_r grows like $\Omega^{1/2}$ but the growth rate $-\alpha_i$ decays, as indicated in Fig. 5. In addition, the results (4.6a-c) have wide application, to most boundary layers and at most x_0's in fact, once account is taken of the factors λ, u_e which were scaled out near the end of Section 3. Comparisons between predictions stemming from (4.6a-c) and experiments are presented earlier in Fig. 4, showing fair agreement. There is similar agreement with computations, although the latter are often for the Orr-Sommerfeld approximation, based on linearising the Navier-Stokes equations about $[u, v] = [\bar{u}(\bar{y}), 0]$ only, which is strictly incorrect at finite *Re* of course.

Finally here, three other items are the following. First, the behaviour near the upper branch of the neutral curve in Fig. 6 is rather similar, related to the high-frequency response in (4.6c). Second, if the velocity profile $\bar{u}(\bar{y})$ is inflectional then the upper branch tends to a nonzero limit at large *Re*.

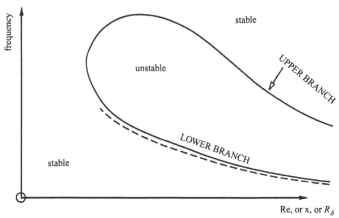

Figure 6: Sketch of the neutral curve of frequency ω versus *Re* or x or R_δ.

Third, other related flows induce various pressure-displacement laws such as

$$P = \pm A, \ -A_X, \ \pm A_{XX} \quad \text{etc.}, \tag{4.7}$$

or mixtures thereof, instead of (3.6e). The examples in (4.7) are for rivers, hypersonic flow, supersonic flow, channel flows, jets, etc..

5   Nonlinear Unsteadiness/Transition.

To recap, we wish to understand (3.6a,c-e), (4.2) in full, given that linear theory yields Tollmien-Schlichting waves (see Section 4). So we ask what do the weakly nonlinear and strongly nonlinear theories imply? Here, broadly, *weak* nonlinearity corresponds to only a small relative change in the mean flow from its original form whereas *strong* nonlinearity corresponds to an $\mathcal{O}(1)$ or larger change. Both fixed-frequency and initial-value problems are of interest now, with weak nonlinearity being based around $\Omega \approx \Omega_0$ or $\Omega \gg 1$ since those are near-neutral values, while strong nonlinearity brings in the whole frequency range in effect. In this section, for unsteady two-dimensional motion still, weakly nonlinear effects are considered first in Sections 5.1 and 5.2. These however are found to have only a little direct relevance to strongly nonlinear effects, which are considered in Sections 5.3–5.5.

5.1   Weak nonlinearity near $\Omega = \Omega_0$

For small disturbance amplitudes h, again guided by Sections 3 and 4, we expect

$$\Omega = \Omega_0 + \mathcal{O}(h^2), \ \alpha = \alpha_0 + \mathcal{O}(h^2), \ E \equiv \exp\left\{i(\alpha_0 X - \Omega_0 T)\right\},$$

$$U = Y + h(U_1 E + c.c.) + h^2 U_2 + h^3 U_3 + \cdots , \tag{5.1}$$

and so on, with the multiple scaling $\partial_X \to \partial_X + h^2 \partial_{\bar{X}} + \cdots$. Here U_2 contains second harmonics $\propto E^{\pm 2}$ and a mean-flow correction $\propto E^0$, U_3 contains $E^{\pm 3}, E^{\pm 1}$, and so on (see references below). Substitution into (3.6a,c-e), (4.2) then leads to the complex amplitude equation

$$\frac{dA_1}{d\bar{X}} = i\alpha_2 A_1 + a_1 \left|A_1\right|^2 A_1, \tag{5.2}$$

in traditional fashion (Stuart [44], Watson [47]) after some working. The coefficient a_1 turns out to be such that supercritical stability holds (Smith [27], the results agreeing with computations by Itoh [19]).

Forced flows add an extra term to (5.2), c.f. (4.4a-d), (4.5) (Hall & Smith [12]). As Ω increases further from Ω_0 computations become necessary. Conlisk, Burggraf & Smith [6] provide numerical travelling-wave solutions which continue on from (5.2). Again, in three-dimensional unsteady flow, (5.2) is replaced by a coupled multi-mode system of amplitude equations.

5.2   Weak nonlinearity for $\Omega \gg 1$

This high-frequency analysis stems from (4.6c) and Fig. 5 which suggest, for large Ω, the multiple scaling

$$\partial_T \to \Omega\partial_{T_0} + \Omega^{1/2}\partial_{T_1} + \cdots , \qquad \partial_X \to \Omega^{1/2}\partial_{X_0} + \partial_{X_1} + \cdots , \qquad (5.3a)$$

and the expansions

$$P = \Omega^{1/2}P_0 + P_1 + \cdots \qquad (P_0 \equiv P_{01}E + c.c.), \qquad (5.3b)$$

where now $E \equiv \exp\{i(X_0 - T_0)\}$. Substitution again into (3.6a,c-e), (4.2) leads, after some working, to the amplitude equation

$$\frac{\partial P_{01}}{\partial T_2} - i\frac{\partial^2 P_{01}}{\partial X_1^2} = \left(\frac{1-i}{2^{1/2}}\right)P_{01} - \frac{5i}{2}|P_{01}|^2 P_{01}, \qquad (5.4)$$

a Ginzburg-Landau form. Computations and analysis of (5.4) for large times T_2 are in Smith [32] and Fig. 7, indicating strong amplitude growth and spreading.

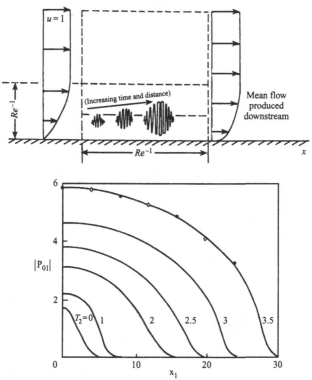

Figure 7: 2D disturbances in boundary layers.

5.3   Strong nonlinearity for $\Omega \gg 1$

Following on from Section 5.2, a new stage arises with higher amplitudes present,

$$P = \Omega \bar{P}_0 + \cdots , \qquad (5.5)$$

and corresponding scales $\Omega^{1/2}, \Omega^{-1/2}$ in A, X and so on. Here in essence the viscous term in (4.2) is neglected, leaving $U \equiv Y + A$, $V \equiv -Y A_X$, so that (4.2) with (3.6e) implies

$$A_T + A A_X = -\frac{1}{\pi} \fint_{-\infty}^{\infty} \frac{A_{ss}(s,T)ds}{(X-s)}. \qquad (5.6)$$

This is the Benjamin-Ono equation for $A(X,T)$, which yields interesting nonlinear travelling-wave and solitary-wave solutions (Zhuk & Ryzhov [50], Smith & Burggraf [31]). Related flows as in (4.7) produce the Burgers and the KdV equations, among others, instead of (5.6). In virtually all cases the solutions indicate a well-behaved regular response for A, P, etc..

On the other hand, a viscous sublayer is necessary nearer the wall since (5.6) is based on an inviscid tangential-flow assumption at $Y = 0+$ (c.f. Sections 2.1 and 2.2). The sublayer, of thickness $\mathcal{O}(\Omega^{-1/2})$ in Y, has its (boundary-layer) motion driven by the unsteady effective slip velocity $u_e \equiv A(X,T)$ and in fact is controlled by (2.8a,b',d,f,g). As implied in Section 2.2, this sublayer cannot stay attached throughout the motion, and indeed

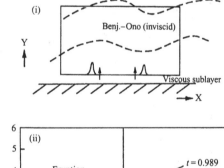

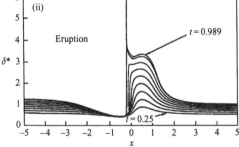

Figure 8: High-frequency strongly nonlinear behaviour (i), suggesting a scale cascade and spiking (ii).

an eruption is likely (Van Dommelen [45], Smith & Burggraf [31]) in which the sublayer thickness δ becomes singular,

$$\delta \to \infty \text{ locally in finite time.} \tag{5.7}$$

The interaction between the eruption (5.7) and the outer flow (c.f. Section 3) is complex (Elliott, Cowley & Smith [8], Peridier, Smith & Walker [24]), especially if subsequent eruptions also arise, and the end-product is still uncertain, although a cascade of scales may occur subsequently (see Fig. 8).

5.4 Strong nonlinearity generally

A clue to the general outcome of the full system (3.6a-e), (4.2) is provided by the special case $P = A$ in (4.7) (Brotherton-Ratcliffe & Smith [4]), for which (5.6) is altered to

$$A_T + (A+1)A_X = 0, \tag{5.8a}$$

or to

$$P_T + PP_X = -\frac{1}{\pi^{1/2}} \int_X^\infty P_s(s,T) \frac{ds}{(s-X)^{1/2}} \tag{5.8b}$$

at modified amplitudes where a viscous contribution appears on the right. The significant feature here is that the solution becomes irregular within a finite time due to the left-hand side inducing a crossing of characteristics effectively, say as $(X,T) \to (X_s, T_s)$. In some detail, locally the X-scale is given by ξ of order one, where

$$(X - X_s) = c(T - T_s) + (T_s - T)^n \xi, \tag{5.9a}$$

with the power $n\ (>1)$ to be found and c constant, while the P solution has

$$P = \pi_0 + (T_s - T)^m \pi_1(\xi) + \cdots \tag{5.9b}$$

with π_0 constant, the power $m(>0)$ again being unknown. Then substitution into (5.8b), which is dominated by its left-hand side, requires $\pi_0 = c$, first, and $m = n - 1$, second. That leaves the nonlinear equation

$$n\xi\pi_1' - (n-1)\pi_1 + \pi_1\pi_1' = 0 \tag{5.9c}$$

for $\pi_1(\xi)$, its general solution being

$$\xi = -\pi_1 - e_1\pi_1^{n/(n-1)} \tag{5.9d}$$

in implicit form, where e_1 is a constant. But $\pi_1(\xi)$ must be smooth and single-valued of course and so $n/(n-1)$ must be an odd integer. Hence the allowable values of n are

$$n = \frac{3}{2}, \frac{5}{4}, \frac{7}{6}, \frac{9}{8}, \cdots. \tag{5.9e}$$

The value associated with smooth initial conditions imposed at some time less than T_s is the first in (5.9e). The singularity or break-up of (5.8a or b) is therefore described by (5.9a-d) with $n = \frac{3}{2}, m = \frac{1}{2}$ in general. The same singularity is then found (see last reference) to apply at finite frequencies, i.e. for the whole system (3.6a,c,d), (4.2) under the interaction law $P = A$.

The above is generalised in Smith [33] to cover all the interaction laws of interest (as in (3.6e), (4.7)), the finite-time break-up then having

$$U = U_0(Y) + (T_s - T)^{1/2} U_1(\xi, Y) + \cdots, \tag{5.10a}$$

$$P = \pi_0 + (T_s - T)^{1/2} \pi_1(\xi) + \cdots, \tag{5.10b}$$

where

$$(X - X_s) = c(T - T_s) + (T_s - T)^{3/2} \xi \tag{5.10c}$$

usually. Substitution into (3.6a,c-e), (4.2) gives the implicit expression

$$\xi = -\pi_1 - \pi_1^3 \tag{5.10d}$$

for the local pressure variation $\pi_1(\xi)$ in normalised form. The same holds for cases (4.7). The break-up (5.10a-d) means that the pressure-gradient solution P_X at the finite time $T = T_s$ develops an inverse-two-thirds singularity at $X = X_s\pm$. Comparisons with computations and experiments are favourable (see Fig. 9 and Peridier, Smith & Walker [24], Smith & Bowles [35]).

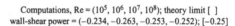

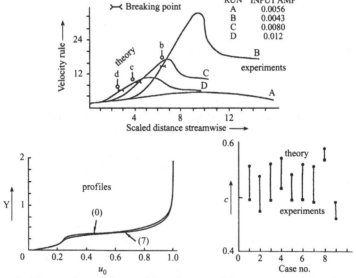

Figure 9: Comparisons of transition theory with computations and experiments.

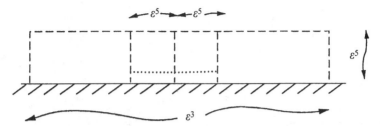

Figure 10: Long-scale/short-scale interaction, plus sublayer eruptions, from Section 5.5.

5.5 What happens next?

Following the singular end of the full interactive stage (3.6a,c-e), (4.2) implied in Section 5.4 new scales must enter play. These shorter scales are examined in the references below. In brief the normal pressure gradient $\partial p/\partial y$ is found to become significant locally in the wall sublayer near $X = X_s$, leading eventually to an extended-KdV equation of the normalised form

$$\tilde{p}_{\tilde{t}} + \tilde{p}\tilde{p}_{\tilde{X}} = \tilde{p}_{\tilde{X}\tilde{X}\tilde{X}} + \tilde{\mu}\tilde{p}_{\tilde{X}} \int_{-\infty}^{\infty} \frac{(\hat{p}_{\tilde{t}} + \hat{p}\hat{p}_{\hat{X}})}{(\tilde{p} - \hat{p})}d\hat{X}, \tag{5.11}$$

controlling the unknown scaled wall pressure $\tilde{p}(\tilde{X}, \tilde{t})$. Here $\tilde{\mu}$ is a constant. Computations and analysis (Hoyle, Smith & Walker [17], He, Walker, Bowles & Smith [16]) then indicate that the system provokes even shorter scales subsequently, for any $\tilde{\mu}$ value. If $\tilde{\mu}$ is zero for instance an ensuing stage results in which new long/short interactions are forced between the interactive-boundary-layer properties in the original $\mathcal{O}(\epsilon^3)$ length scale and the Euler properties in a new $\mathcal{O}(\epsilon^5)$ length scale, as summarised in Fig. 10. Again, however, sublayer eruptions as described in Section 5.3 are then likely, leading on to another cascade process of interest.

6 Three-Dimensional Nonlinearity

While acknowledging the importance of three-dimensionality both theoretically and practically, we have to be especially brief here.

There are clearly many solution paths in three-dimensional unsteady flows. Some originate from the extension of (3.6a,c-e), (4.2) to three spatial dimensions:

$$U_X + V_Y + W_Z = 0, \tag{6.1a}$$

$$U_T + UU_X + VU_Y + WU_Z = -P_X(X, Z, T) + U_{YY}, \tag{6.1b}$$

$$W_T + UW_X + VW_Y + WW_Z = -P_Z(X, Z, T) + W_{YY}, \tag{6.1c}$$

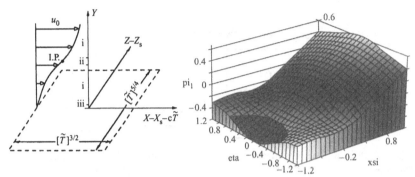

Figure 11: A 3D transition path, from Hoyle & Smith [18]; see others in Section 6.

with $U = V = W = 0$ at $Y = 0$, $W \to 0$ as $Y \to \infty$, and with (3.6e) (or (4.7)) enlarged appropriately. Then various weak and strong nonlinearities apply, either analogous with those of Sections 5.1–5.4 (e.g. Hoyle & Smith [18], Fig. 11) or completely fresh, i.e. not observed in two dimensions. Examples of the latter are resonant triads (see below) and vortex/wave interactions (Hall & Smith [14, 15], Blennerhassett & Smith [3]). The triads are inferred from the extension of (4.6c) to three dimensions, namely

$$\alpha(\alpha^2 + \beta^2)^{1/2} = \Omega, \tag{6.2a}$$

for a given spanwise Z-wavenumber β. Here (6.2a) implies that three waves are possible of the form

$$(\alpha, \beta, \Omega) = (\alpha_1/2, \beta_1, \Omega_1/2) \text{ and } (\alpha_1/2, -\beta_1, \Omega_1/2) \text{ and } (\alpha_1, 0, \Omega_1), \tag{6.2b}$$

provided that $\beta_1 = 3^{1/2}\alpha_1/2$ and $\alpha_1^2 = \Omega_1$. Thus these three waves can co-exist and interact weakly nonlinearly only for a critical wave angle of 60°, with the first and second waves combining to force the third and so on. The above forms the basis of several papers (Craik [7], Smith & Stewart [37], Wu [48]). In contrast, vortex/wave interaction is more general since it needs only two waves, say

$$(\alpha, \beta, \Omega) \text{ and } (\alpha, -\beta, \Omega), \tag{6.2c}$$

as those combine to force a vortex component $(0, 2\beta, 0)$. This type of interaction, occurring both near $\Omega = \Omega_0$ and for large Ω, is possible for any nonzero wave angle.

Vortex/wave nonlinear interactions are of particular interest as they arise first at remarkably low amplitudes, usually lower than for nonlinear interactions in two dimensions. Three examples that emerge from (6.1a-c) essentially are in Smith & Walton [34], Stewart & Smith [38], Smith & Bowles [35]. The first has the coupled equations

$$(c_1 + c_2 X + c_3 \lambda_3)\tilde{P} + c_4 \tilde{P}_{ZZ} + c_5 \tilde{P} \left|\tilde{P}\right|^2 = 0, \tag{6.3a}$$

$$\lambda_3 = \partial_Z^2 \int_{X_0}^X \left| \tilde{P} \right|^2 (\xi, Z) \left\{ c_6 + c_7 \ell n(X - \xi) \right\} (X - \xi)^{-1/3} d\xi, \qquad (6.3b)$$

controlling the scaled wave pressure $\tilde{P}(X, Z)$ and the scaled vortex shear $\lambda_3(X, Z)$. The c_n are constants. The solution properties would be benign in two dimensions where ∂_Z is identically zero. By contrast, in three dimensions the system (6.3a,b) exhibits strong secondary instability. Similar phenomena occur in the other references above and elsewhere, accompanied by finite-distance blow-ups in the disturbance amplitudes. These blow-ups would not arise in two dimensions. Strong nonlinearity has also been considered. The theoretical properties here are in fair agreement with computations and experiments (see the references above).

Miscellaneous points in three dimensions are as follows: secondary instability is quite typical, c.f. the Benjamin-Feir [1] mechanism; three-dimensional nonlinear effects are often more "dangerous" than are two-dimensional ones, as hinted earlier; vortex/wave interactions occur in numerous forms, for example coupling partial-differential and ordinary-differential equations for the vortex and wave respectively (as in references above); and there are multiple modes present (Hall & Smith [13]).

Finally, as indicated above, this article has focused mostly on nonlinear effects originating rationally from triple-deck theory, for convenience. There are many others, for instance involving upper-branch or inflectional instabilities (Section 4) initially, e.g. Smith, Brown & Brown [36], Wu, Lee & Cowley [49], Goldstein & Wundrow [9], Brown & Smith [5]. The whole lot shows a rich variety of three-dimensional interactions being available.

References

[1] Benjamin, T.B. and Feir, J.E., The disintegration of wave trains on deep water. Part 1. Theory, *J. Fluid Mech.* **27**, 417-430, 1967.

[2] Bertolotti, F.P., Herbert, Th. and Spalart, P.R., Linear and nonlinear stability of the Blasius boundary layer, *J. Fluid Mech.* **242**, 442-474, 1992.

[3] Blennerhassett, P. and Smith, F.T., Nonlinear interaction of oblique three-dimensional Tollmien-Schlichting waves and vortices in channel flows and boundary layers, *Proc. Roy. Soc. Lond.* A **436**, 585-602, 1992.

[4] Brotherton-Ratcliffe, R.V. and Smith, F.T., Complete breakdown of an unsteady interactive boundary layer (over a surface distortion or in a liquid layer), *Mathematika* **34**, 86-100, 1987.

[5] Brown, S.N. and Smith, F.T., On vortex/wave interactions. Part I: Non-symmetrical input and cross-flow in boundary layers, *J. Fluid Mech.*, in press, 1996.

[6] Conlisk, A.T., Burggraf, O.R. and Smith, F.T., Nonlinear neutral modes in the Blasius boundary layers, *Forum on Unsteady Flow Separation* **32**, 119-121, 1987.

[7] Craik, A.D.D., Evolution in space and time of resonant wave triads. II A class of exact solutions, *Proc. Roy. Soc. A* **363**, 257-269, 1978.

[8] Elliott, J.W., Cowley, S.J. and Smith, F.T., Breakdown of boundary layers: (i) on moving surfaces; (ii) in semi-similar unsteady flow; (iii) in fully unsteady flow, *Geophys. Astrophys. Fluid Dyn.* **25**, 77-138, 1983.

[9] Goldstein, M.E. and Wundrow, D.W., Interaction of oblique instability waves with weak streamwise vortices, *J. Fluid Mech.* **284**, 377-407, 1995.

[10] Goldstein, S., Concerning some solutions of the boundary-layer equations in hydrodynamics, *Proc. Camb. Phil. Soc.* **26**, 1-30, 1930.

[11] Goldstein, S., On laminar boundary layer flow near a point of separation, *Quart. J. Mech. Appl. Math.* **1**, 43-69, 1948.

[12] Hall, P. and Smith, F.T., A suggested mechanism for nonlinear wall roughness effects on stream flow stability, *Stud. Appl. Math.* **66**, 241-265, 1982.

[13] Hall, P. and Smith, F.T., On the effects of nonparallelism, three-dimensionality and mode interaction in nonlinear boundary-layer stability, *Stud. Appl. Math.* **70**, 91-120, 1984.

[14] Hall, P. and Smith, F.T., The nonlinear interaction of Tollmien-Schlichting waves and Taylor-Gortler vortices in curved channel flows, *Proc. Roy. Soc. A* **417**, 255-282, 1988.

[15] Hall, P. and Smith, F.T., On strongly nonlinear vortex/wave interactions in boundary-layer transition, *J. Fluid Mech.* **227**, 641-666, 1991.

[16] He, J., Walker, J.D.A., Bowles, R.I. and Smith, F.T., Short-scale break-up in unsteady interactive boundary layers: local development of normal pressure gradients, submitted to *J. Fluid Mech.*, 1995.

[17] Hoyle, J.M., Smith, F.T. and Walker, J.D.A., On sublayer eruption and vortex formation, *Comp. Phys. Comms.* **65**, 151-157, 1991.

[18] Hoyle, J.M. and Smith, F.T., On finite-time break-up in three-dimensional unsteady interacting boundary layers, *Proc. Roy. Soc. Lond. A* **447**, 467-492, 1994.

[19] Itoh, N., *Trans. Jap. Soc. Aeronaut. Space Sci.* **17**, 174-191, 1974.

[20] Jones, C.W. and Watson, E.J., Two-dimensional boundary layers, Ch. V of *Laminar Boundary Layers*, ed. L. Rosenhead, OUP, 1963.

[21] Kleiser, L. and Zang, T.A., Numerical simulation of transition in wall-bounded shear flows, *Ann. Rev. Fluid Mech.* **23**, 495-537, 1991.

[22] Messiter, A.F., Boundary layer flow near the trailing edge of a flat plate, *SIAM J. Appl. Math.* **18**, 241-257, 1970.

[23] Messiter, A.F., Boundary-layer interaction theory, *Trans. ASME J. Appl. Mech.* **50**, 1104-1113, 1983.

[24] Peridier, V.J., Smith, F.T. and Walker, J.D.A., Vortex induced boundary-layer separation. Part 1. The unsteady limit problem $Re \to \infty$. Part 2. Unsteady interacting boundary-layer theory, *J. Fluid Mech.* **232**, 99-131, 133-165, 1991.

[25] Smith, F.T., Laminar flow over a small hump on a flat plate, *J. Fluid Mech.* **57**, 803-824, 1973.

[26] Smith, F.T., On the non-parallel flow stability of the Blasius boundary layer, *Proc. Roy. Soc. A* **366**, 91-109, 1979.

[27] Smith, F.T., Nonlinear stability of boundary layers for disturbances of various sizes, *Proc. Roy. Soc. A* **368**, 573-589, 1979 (and corrections **371**, 439-440, 1980).

[28] Smith, F.T. and Daniels, P.G., Removal of Goldstein's singularity at separation in flow past obstacles in wall layers, *J. Fluid Mech.* **110**, 1-38, 1981.

[29] Smith, F.T., On the high Reynolds number theory of laminar flows, *IMA J. Appl. Math.* **28**, 207-281, 1982.

[30] Smith, F.T., Papageorgiou, D. and Elliott, J.W., An alternative approach to linear and nonlinear stability calculations at finite Reynolds numbers, *J. Fluid Mech.* **146**, 313-330, 1984.

[31] Smith, F.T. and Burggraf, O.R., On the development of large-sized short-scaled disturbances in boundary layers, *Proc. Roy. Soc. Lond. A* **399**, 25-55, 1985.

[32] Smith, F.T., Steady and unsteady boundary layer separation, *Ann. Rev. Fluid. Mech.* **18**, 197-220, 1986.

[33] Smith, F.T., Finite-time break-up can occur in any unsteady interactive boundary layer, *Mathematika* **35**, 256-273, 1988.

[34] Smith, F.T. and Walton, A.G., Nonlinear interaction of near-planar TS waves and longitudinal vortices in boundary-layer transition, *Mathematika* **36**, 262-289, 1989.

[35] Smith, F.T. and Bowles, R.I., Transition theory and experimental comparisons on (I) amplifications into streets and (II) a strongly nonlinear break-up criterion, *Proc. Roy. Soc. A* **439**, 163-175, 1992.

[36] Smith, F.T., Brown, S.N. and Brown, P.G., Initiation of three-dimensional nonlinear transition paths from an inflectional profile, *Eur. J. Mech., B/Fluids* **12**, 447-473, 1993.

[37] Stewart, P.A. and Smith, F.T., Three-dimensional instabilities in steady and unsteady non-parallel boundary layers, including effects of Tollmien-Schlichting disturbances and cross flow, *Proc. Roy. Soc. A* **409**, 229-248, 1987.

[38] Stewart, P.A. and Smith, F.T., Three dimensional nonlinear blow-up from a nearly planar initial disturbance in boundary-layer transition: theory and experimental comparisons, *J. Fluid Mech.* **244**, 79-100, 1992.

[39] Stewartson, K., On the flow near the trailing edge of a flat plate - II, *Mathematika* **16**, 106-121, 1969.

[40] Stewartson, K. and Williams, P.G., Self-induced separation, *Proc. Roy. Soc. A* **312**, 181-206, 1969.

[41] Stewartson, K., Is the singularity at separation removable?, *J. Fluid Mech.* **44**, 347-364, 1970.

[42] Stewartson, K., On laminar boundary layers near corners, *Quart. J. Mech. Appl. Math.* **23**, 137-152, 1970 (and corrections **24**, 387-389, 1970).

[43] Stewartson, K. and Williams, P.G., Self-induced separation II, *Mathematika* **20**, 98-108, 1973.

[44] Stuart, J.T., On the non-linear mechanics of wave disturbances in stable and unstable parallel flows. Part 1. The basic behaviour in plane Poiseuille flow, *J. Fluid Mech.* **9**, 353-370, 1960.

[45] Van Dommelen, L.L., Unsteady boundary-layer separation, PhD thesis, Cornell University, 1981.

[46] Van Dyke, M., *An Album of Fluid Motion*, Parabolic Press, 1982.

[47] Watson, J., On the non-linear mechanics of waves disturbances in stable and unstable parallel flows. Part 2. The development of a solution for plane Poiseuille flow and for plane Couette flow, *J. Fluid Mech.* **9**, 371-389, 1960.

[48] Wu, X., The nonlinear evolution of high-frequency resonant-triad waves in an oscillatory Stokes-layer at high Reynolds number, *J. Fluid Mech.* **245**, 553-597, 1992.

[49] Wu, X., Lee, S.S. and Cowley, S.J., On the weakly nonlinear three-dimensional instability of shear layers to pairs of oblique waves: the Stokes layer as a paradigm, *J. Fluid Mech.* **253**, 681-721, 1993.

[50] Zhuk, V.I. and Ryzhov, O.S., Locally inviscid perturbations in a boundary layer with self-induced pressure, *Soviet Phys. Dokl.* **27**, 177-179, 1982.

Recent Developments on Wave Instability

T.J. Bridges
Department of Mathematical and Computing Sciences,
University of Surrey,
Guildford

1 Introduction

Waves appear in many physical problems such as oceanic flow, the nervous system, atmospheric flows, shear flows, wave guides, acoustics, gas dynamics and many other areas. When modelling wave propagation using partial differential equations a natural question that must be addressed is the stability of waves since the stability (or instability) of a wave in a model equation will strongly suggest the appearance (or non-appearance) of such waves in nature.

When considering the linear stability of solutions of time-dependent ordinary differential equations (ODE's), partial differential equations (PDE's) and wave systems it is often considerably easier to prove instability than stability. To prove linear instability it is sufficient to exhibit a particular solution with an unstable linear stability exponent. Recall the situation for ODE's; let $u \in \mathbf{R}^n$ and consider the ODE

$$u_t = F(u), \qquad u(t_0) = u_0 \in \mathbf{R}^n, \qquad (1.1)$$

for some smooth function $F : \mathbf{R}^n \to \mathbf{R}^n$. Suppose $\hat{u} \in \mathbf{R}^n$ satisfies $F(\hat{u}) = 0$; an equilibrium solution. The linear stability problem for $\hat{u}$ is obtained by linearising (1.1) about $\hat{u}$ giving

$$u_t = \mathbf{A}u, \qquad \text{where } \mathbf{A} = DF(\hat{u}) \qquad (1.2)$$

and $DF(\hat{u})$ is the Jacobian of $F(u)$ evaluated at $\hat{u}$. If there exists an eigenvalue λ of the matrix $\mathbf{A}$ with $\operatorname{Re}(\lambda) > 0$ then we say that the basic state $\hat{u}$ is linearly unstable, since, if $\lambda \in \mathbf{C}$ is an eigenvalue of $\mathbf{A}$ with eigenfunction ξ then $\mathbf{A}\xi = \lambda\xi$ in which case $u(t) = \operatorname{Re}(\xi e^{\lambda t})$ is an exponentially growing solution of (1.2).

Note that no further information about the linear operator $\mathbf{A}$ or the other eigenvalues of $\mathbf{A}$ is required. The existence of at least one unstable eigenvalue is sufficient. In the finite-dimensional setting (i.e. the analysis of ODE's) more complete information about the spectrum of $\mathbf{A}$ is often available. However the idea that existence of a single unstable eigenvalue is sufficient to conclude

linear instability is a powerful one and especially useful for studying the instability of wave solutions of PDE's where a full understanding of the spectrum
of the linearised system is often intractable.

To appreciate the difficulties that arise when studying the linear stability
of waves consider the following model problem

$$u_t = u_{xx} + V'(u), \quad -\infty < x < \infty, \quad t > 0, \tag{1.3}$$

for a scalar-valued function $u(x,t)$ where $V(u)$ is some smooth function of
u. Suppose there exists a spatially periodic "wave" solution $\hat{u}(x)$ of (1.3)
satisfying

$$\frac{d^2}{dx^2}\hat{u}(x) + V'(\hat{u}(x)) = 0 \quad \text{with} \quad \hat{u}(x+L) = \hat{u}(x), \tag{1.4}$$

for all $x \in \mathbf{R}$ and some positive number L. It is straightforward to establish sufficient conditions for the existence of such states using phase plane
techniques.

The linear stability problem for $\hat{u}(x)$ is formulated by linearising (1.3)
about the solution (1.4). With $u(x,t) = \hat{u}(x) + v(x,t)$, the perturbation
function $v(x,t)$ satisfies

$$v_t = v_{xx} + V''(\hat{u}(x))v$$

or

$$v_t = \mathbf{A}v \quad \text{with} \quad \mathbf{A} = \frac{d^2}{dx^2} + V''(\hat{u}(x)). \tag{1.5}$$

The situation is, abstractly, similar to the ODE case in (1.2); if there exists
an eigenvalue $\lambda \in \mathbf{C}$ of the operator $\mathbf{A}$ with $\text{Re}(\lambda) > 0$ we say that the
basic state $\hat{u}(x)$ is linearly unstable. However, the study of the spectrum of
differential operators such as $\mathbf{A}$ in (1.5) is much more difficult.

The instability question for (1.5) is formulated as follows. Suppose that
there exists a function $v(x)$ and $\lambda \in \mathbf{C}$ with $v(x)$ bounded for all $x \in \mathbf{R}$
and satisfying $\mathbf{A}v = \lambda v$ with $\text{Re}(\lambda) > 0$, then the basic state $\hat{u}(x)$ is linearly
unstable.

On the other hand, to prove that $\hat{u}(x)$ is linearly stable, it is necessary
to show that *every bounded solution* of $\mathbf{A}v = \lambda v$, with $\mathbf{A}$ defined in (1.5),
corresponds to a value of λ with $\text{Re}(\lambda) < 0$. Note that this does not preclude
the existence of solutions of $\mathbf{A}v = \lambda v$ with $\text{Re}(\lambda) > 0$ but, when the basic
state is linearly stable, such solutions will not be bounded for all $x \in \mathbf{R}$.

Clearly the linear stability question is much more difficult and in many
cases is intractable. In this article we review recent developments of methods
for proving sufficient conditions for the existence of unstable eigenvalues; a
problem which is tractable and leads to new results on wave dynamics in open
systems.

In Section 2 the model problem (1.3) is considered in more detail and, with a combination of Floquet theory and a geometric analysis of the basic state, a criterion for the existence of unstable eigenvalues is presented. Although the model problem (1.3) is fairly simple, the formulation and proof are a prototype for more complex wave problems. In Section 3 the formulation of Section 2 is abstracted and then extended to the instability problem for spatially quasiperiodic waves. Here there are interesting connections with the geometry of invariant tori. Stability of quasiperiodic waves is especially difficult to analyse since Floquet's Theorem is not in general applicable.

The instability problem for a class of dispersive wave equations is formulated in Section 4. It is shown that the geometric instability criterion for periodic travelling waves is similar to that for stationary quasiperiodic waves. Moreover there are interesting connections with the Whitham modulation theory; the instability criterion provides a rigorous proof of the instability predicted by the Whitham modulation equation. In Section 5 a sketch of some other recent developments is presented.

Background material on the theory of nonlinear waves can be found in the book by Whitham [16]. Recent books which devote sections to the stability of nonlinear waves are Infeld & Rowlands [13] and van Groesen & de Jager [12].

2   Spatially Periodic States of $u_t = u_{xx} + V'(u)$

In this section a sufficient geometric condition for linear instability of periodic states of the model problem (1.3) is presented. It is a well known "Folk Theorem" that *all* spatially periodic states of (1.3) are unstable; one can prove this by applying the theory of Hill's equation (e.g. Lemma 4.5 on page 59 of Magnus & Winkler [14]) to $\mathbf{A}v = \lambda v$ in (1.5) which is equivalent to Hill's equation (c.f. equation (2.1) on page 11 of Magnus & Winkler [14]). However, the purpose of this section is to introduce, in the simplest possible setting, a geometric criterion for instability, which, while not particularly useful for this simple model problem, extends to, and is extremely useful for, more complicated PDE's with wave solutions.

The first step is to transform (1.3) into a form more amenable to analysis. Let $v = u_x$ and then (1.3) can be recast as

$$
\begin{aligned}
u_x &= v \\
-u_t + v_x &= -V'(u)
\end{aligned}
$$

or

$$
\mathbf{M}Z_t + Z_x = F(Z), \qquad Z = \begin{bmatrix} u \\ v \end{bmatrix} \equiv \begin{bmatrix} u \\ u_x \end{bmatrix}, \tag{2.1}
$$

with

$$\mathbf{M} = \begin{bmatrix} 0 & 0 \\ -1 & 0 \end{bmatrix} \quad \text{and} \quad F(Z) = \begin{bmatrix} v \\ -V'(u) \end{bmatrix} = \mathbf{J}\nabla S(Z), \qquad (2.2)$$

where

$$\mathbf{J} = \begin{bmatrix} 0 & 1 \\ -1 & 0 \end{bmatrix} \quad \text{and} \quad S(Z) = \tfrac{1}{2}v^2 + V(u). \qquad (2.3)$$

The advantage of writing (1.3) in the form (2.1) is that time independent states correspond to orbits of the ordinary differential equation $Z_x = F(Z)$. Note however that there is further structure here in that $F(Z) = \mathbf{J}\nabla S(Z)$; in particular, time-independent states are orbits of a single degree of freedom Hamiltonian system

$$Z_x = \mathbf{J}\nabla S(Z) \qquad \text{for } Z \in \mathcal{M} = \mathbf{R}^2, \qquad (2.4)$$

with Hamiltonian function $S(Z)$. It follows immediately from (2.4) that $S(Z)$ is an absolute spatial invariant since

$$\frac{d}{dx}S(Z) = \langle \nabla S(Z), Z_x \rangle = \langle \nabla S(Z), \mathbf{J}\nabla S(Z) \rangle = 0,$$

since $\mathbf{J}$ is skew-symmetric (where $\langle \cdot, \cdot \rangle$ is the standard inner product on $\mathbf{R}^2$).

Suppose that there exists a spatially periodic solution of (2.4), denoted $\widehat{Z}(x; s)$, satisfying

$$\widehat{Z}_x = \mathbf{J}\nabla S(\widehat{Z}), \qquad \widehat{Z}(x + L(s); s) = \widehat{Z}(x; s), \qquad (2.5)$$

where $s \in \mathbf{R}$ parametrises the branch of periodic states. Since $S(\widehat{Z}) = s$ is a spatial invariant it is natural to parametrise the branch of solutions by the value of $S(\widehat{Z})$, the "spatial energy". Note that $S(\widehat{Z})$ is a functional and $s \in \mathbf{R}$ is the value of the functional. This parametrisation leads to

$$1 = \frac{dS}{ds} = \langle \nabla S(\widehat{Z}), \widehat{Z}_s \rangle = \langle \mathbf{J}^{-1}\widehat{Z}_x, \widehat{Z}_s \rangle = \langle \widehat{Z}_x, \mathbf{J}\widehat{Z}_s \rangle, \qquad (2.6)$$

an identity which will be useful later. What we will prove is the following.

Theorem 1 *Let $\widehat{Z}(x; s)$ be an $L(s)$-periodic solution of (2.4) and suppose that*

$$L'(s) > 0. \qquad (2.7)$$

Then $\widehat{Z}(x; s)$ is linearly unstable and there exists a spatially bounded solution of the linear stability problem with $\mathrm{Re}(\lambda) > 0$.

Proof: To prove this result we have to study the linear stability problem (1.5) and show that (2.7) implies the existence of a bounded state with an unstable eigenvalue. Linearise (2.1) about the state $\widehat{Z}$; letting $Z(x,t) = \widehat{Z}(x;s) + \widehat{U}(x,t)$ gives

$$\mathbf{M}\widehat{U}_t + \widehat{U}_x = \mathbf{J}D^2S(\widehat{Z})\widehat{U}, \tag{2.8}$$

where $D^2S(\widehat{Z})$ is the second derivative of $S(Z)$ evaluated at $\widehat{Z}$. Since $\widehat{Z}$ is independent of t we take $\widehat{U}(x,t) = e^{\lambda t}U(x)$ with U satisfying

$$U_x = [\mathbf{J}D^2S(\widehat{Z}) - \lambda\mathbf{M}]U \tag{2.9}$$

and the linear instability question is as follows. If there exists a bounded state of (2.9) with $\text{Re}(\lambda) > 0$ then the basic state is linearly unstable. What we need to prove is that (2.7) implies the existence of an unstable solution.

The proof is completed in two steps. First, since $\widehat{Z}$ is L-periodic the coefficients in the equation are L-periodic and hence Floquet's theorem applies: every bounded solution of (2.9) is of the form $U(x) = \mathcal{V}(x)e^{i\alpha x}$ where $\alpha \in \mathbf{R}$ and $\mathcal{V}(x)$ is L-periodic and satisfies $\Phi(\mathcal{V}; \lambda, \alpha) = 0$ where

$$\Phi(\mathcal{V}; \lambda, \alpha) \equiv \mathcal{V}_x - [\mathbf{J}D^2S(\widehat{Z}) - \lambda\mathbf{M} - i\alpha\mathbf{I}]\mathcal{V}. \tag{2.10}$$

Let $\mathcal{C}_L^j$ for $j = 0,1$ be the space of L-periodic, j-times continuously differentiable functions. Then (2.10) can be treated as an operator equation $\Phi : \mathcal{C}_L^1 \times \mathbf{C} \times \mathbf{R} \to \mathcal{C}_L^0$ and

$$\Phi(\mathcal{V}; 0,0) \equiv \mathbf{L}_0\mathcal{V} = \frac{d\mathcal{V}}{dx} - \mathbf{J}D^2S(\widehat{Z})\mathcal{V}. \tag{2.11}$$

The function $\widehat{Z}_x \in \text{Ker}(\mathbf{L}_0)$ since differentiating (2.4) results in

$$(\widehat{Z}_x)_x = \mathbf{J}D^2S(\widehat{Z})\widehat{Z}_x \quad \text{or} \quad \mathbf{L}_0\widehat{Z}_x = 0.$$

We leave it as an exercise to prove that $\text{Ker}(\mathbf{L}_0)|_{\mathcal{C}_L^1} = \{\widehat{Z}_x\}$ when $L'(s) \neq 0$; that is, the Kernel on $\mathcal{C}_L^1$ is not larger.

The second step in the proof is to apply the Lyapunov-Schmidt reduction to (2.10). Introduce the L_2 inner product $[a, b] = \frac{1}{L}\int_0^L \langle a, b \rangle \, dx$ for L-periodic $\mathbf{C}^2$-valued continuous functions. It follows from Fredholm theory that

$$\text{Range}(\mathbf{L}_0) = \{\mathcal{V} \in \mathcal{C}_L^0 \; : \; [\mathbf{J}\widehat{Z}_x, \mathcal{V}] = 0\}. \tag{2.12}$$

Therefore using the projection $\mathbf{E} : \mathcal{C}_L^0 \to \text{Range}(\mathbf{L}_0)$, we have the Lyapunov-Schmidt splitting

$$\mathbf{E} \cdot \Phi(\mathcal{V}; \lambda, \alpha) = 0 \quad \text{and} \quad (\mathbf{I} - \mathbf{E}) \cdot \Phi(\mathcal{V}; \lambda, \alpha) = 0, \tag{2.13}$$

where $\mathcal{V} = \delta\widehat{Z}_x + W$ with $W|_{\lambda=\alpha=0} = 0$ and $\delta \in \mathbf{C}$. In the first equation of (2.13) the operator $\mathbf{E} \cdot \mathbf{L}_0$ is invertible and therefore, by the Implicit Function Theorem for Banach spaces, it is uniquely solvable for $W = W(\cdot, \lambda, \alpha)$ when

$|\lambda|^2 + |\alpha|^2$ is sufficiently small. Substitution of the function W into the second equation of (2.13) leads to the "dispersion relation" for $|\lambda|^2 + |\alpha|^2$ sufficiently small (setting $\delta = 1$ for simplicity)

$$\Delta(\lambda, \alpha) \equiv [\mathbf{J}\widehat{Z}_x, \Phi(\widehat{Z}_x + W; \lambda, \alpha)]$$
$$= a_0\lambda + a_1\alpha + a_2\alpha^2 + \mathcal{O}(|\lambda|^2, |\lambda||\alpha|, |\alpha|^3)$$

for some real numbers a_0, a_1 and a_2. We leave it as an exercise to prove that

$$a_0 > 0,$$
$$a_1 = -i[\widehat{Z}_x, \mathbf{J}\widehat{Z}_x] = 0,$$
$$a_2 = -2i[\widehat{Z}_x, \mathbf{J}W_\alpha]\Big|_{\lambda=\alpha=0}$$

Now $W_\alpha^o \equiv W_\alpha|_{\lambda=\alpha=0}$ satisfies $\mathbf{E} \cdot \mathbf{L}_0 W_\alpha^o = -i\widehat{Z}_x$ and so

$$W_\alpha^o = -i\left(x\widehat{Z}_x + \frac{L(s)}{L'(s)}\widehat{Z}_s\right) + c\widehat{Z}_x,$$

for some $c \in \mathbf{C}$. Thus,

$$a_2 = -2i\left[\widehat{Z}_x, \mathbf{J}\left(-i\frac{L(s)}{L'(s)}\right)\widehat{Z}_s\right]$$
$$= -2\frac{L(s)}{L'(s)}\left[\widehat{Z}_x, \mathbf{J}\widehat{Z}_s\right]$$
$$= -2\frac{L(s)}{L'(s)} \qquad \text{(using (2.6))}.$$

It follows that if $L'(s) > 0$ then $\Delta(\lambda, \alpha) = 0$ has a solution

$$\lambda = \text{sgn}(L'(s))\left|\frac{a_2}{a_0}\right|\alpha^2 + \mathcal{O}(|\alpha|^3) \qquad \text{for } |\lambda|^2 + |\alpha|^2 \text{ sufficiently small,}$$

and hence there exists an unstable solution of the linear stability problem. ∎

The unstable wave is of the form $\widehat{U}(x, t) = e^{(\lambda t + i\alpha x)}\mathcal{V}(x)$ where $\mathcal{V}(x)$ is L-periodic. Since $\alpha \neq 0$ there are two wavenumbers in space: $\frac{2\pi}{L}$ and α. In general the unstable wave will be quasiperiodic in space. This is one of the reasons that studying the initial-value problem directly is so difficult. Perturbations with $|\alpha| \ll 1$ are known as sideband perturbations.

There are many other ways to prove Theorem 1 and some have greater generality. A promising method is to use centre-manifold theory on the spatial part of the linear stability problem, a method which carries over to the case where the time-independent part is an elliptic operator. In a centre-manifold framework, the dynamics around the spatially periodic state can also be studied.

3 Invariant Manifolds, Parameters and Eigenvalues

It is useful to abstract the result in Section 2 in order to see which aspects of the analysis can be generalised. First, by writing the PDE as a first order system, wave solutions, either time-independent or steady relative to a moving frame, correspond to *invariant manifolds* of an ODE. For the problem in Section 2, the invariant manifold is a periodic orbit or closed loop in the (spatial) phase space.

Invariant manifolds generally live in multi-parameter families. For example the periodic orbits in Section 2 exist in a one-parameter family parametrised by s, the value of the spatial Hamiltonian function. Therefore there exist two sets of tangent vectors: a set of tangent vectors on the invariant manifold and a set of tangent vectors on parameter space. For the example in Section 2 the two tangent vectors are $\widehat{Z}_x$ and $\widehat{Z}_s$ respectively and these two tangent vectors are used to construct, rigorously, an unstable eigenfunction of the linear stability problem. Note that the tangent vector $\widehat{Z}_s$ was used at the end of the proof of Theorem 1 when constructing W_α^o.

The condition for existence of the unstable eigenfunction uses the geometric condition $L'(s) > 0$. The function $L(s)$ relates two distinct properties of the periodic orbit. In the phase space $L(s)$ for fixed s is the "time" required to traverse the periodic orbit. The time required to traverse the periodic orbit is distinct from the shape or size of the orbit in the phase space, which we can take to be determined by the value of s. In other words $L'(s)$ is the rate of change of the dynamic property with respect to the kinematic or shape property of the periodic orbit.

With the above summary of the abstract framework behind the theory in Section 2 it is useful to generalise it to other problems. A natural generalisation is to consider spatially quasiperiodic waves. For example consider the cubic Ginzburg Landau (GL) equation with real coefficients (c.f. Bridges & Rowlands [9])

$$\frac{\partial \phi}{\partial t} = \frac{\partial^2 \phi}{\partial x^2} + \phi - |\phi|^2 \phi, \qquad (3.1)$$

where $\phi(x, t)$ is complex valued. It is straightforward to write this equation in the form

$$\mathbf{M} Z_t + Z_x = F(Z)$$

by taking

$$Z = \begin{bmatrix} \mathrm{Re}(\phi) \\ \mathrm{Im}(\phi) \\ \mathrm{Re}(\phi_x) \\ \mathrm{Im}(\phi_x) \end{bmatrix}$$

Suppose that the ODE $Z_x = F(Z)$ has an invariant two-torus. (It is straightforward to prove this for the cubic GL equation since the steady part is an integrable ODE.) The 2-torus is a two-dimensional invariant manifold with a two-dimensional tangent space. Moreover such states exist in a two-parameter family. If the flow on the 2-torus is quasiperiodic then there are two wavenumbers k_1 and k_2 which depend on two parameters I_1 and I_2 which could represent values of the invariants of the integrable ODE. (The precise definition of these parameters for the GL equation is given in Bridges & Rowlands [9].) Therefore the natural generalisation of the Jacobian map $L'(s)$ is

$$J_k = \left[\begin{array}{cc} \frac{\partial k_1}{\partial I_1} & \frac{\partial k_1}{\partial I_2} \\ \frac{\partial k_2}{\partial I_1} & \frac{\partial k_2}{\partial I_2} \end{array} \right]$$

and one might conjecture that this matrix could form the basis for an instability criterion for spatially quasiperiodic waves. In fact Bridges & Rowlands [9, Theorem 1], prove that if $\det(J_k) > 0$ then the basic spatially quasiperiodic state is linearly unstable in the full problem. The proof follows the same lines as that given in Section 2 except that the kernel of $\mathbf{L}_0$ is two-dimensional and there are two tangent vectors on parameter space. Note that J_k is the "KAM matrix" that is fundamental to the persistence problem for invariant tori (c.f. Arnold [1, Appendix 8]). The above theory has been applied to the spatially quasiperiodic states that arise in the rotating channel problem (Bridges & Cooper [7]).

4   Instability of Periodic Travelling Waves

Surprisingly the instability problem for periodic travelling waves has a structure similar to that for spatially quasiperiodic waves. We will show that an instability criterion similar to that given at the end of Section 3 arises also in the case of periodic travelling waves. For example consider the system

$$u_{tt} - u_{xx} + V'(u) = 0, \qquad (x, t) \in \mathbf{R} \times \mathbf{R}, \qquad (4.1)$$

where u is scalar-valued and $V(u)$ is some smooth function of u, which can be written as a first-order system

$$\mathbf{M}Z_t + \mathbf{K}Z_x = \nabla S(Z), \qquad Z = \left[\begin{array}{c} Z_1 \\ Z_2 \\ Z_3 \end{array} \right] \equiv \left[\begin{array}{c} u \\ u_t \\ u_x \end{array} \right], \qquad (4.2)$$

by taking

$$\mathbf{M} = \left[\begin{array}{ccc} 0 & -1 & 0 \\ 1 & 0 & 0 \\ 0 & 0 & 0 \end{array} \right], \qquad \mathbf{K} = \left[\begin{array}{ccc} 0 & 0 & 1 \\ 0 & 0 & 0 \\ -1 & 0 & 0 \end{array} \right]$$

and

$$S(Z) = \tfrac{1}{2}(Z_2^2 - Z_3^2) + V(Z_1).$$

There are several advantages of reformulating (4.1) as the first order system (4.2), the most important of which is that it is a suitable framework for proving results on the instability of travelling wave solutions of (4.1) (c.f. Bridges [6] for further details of this multi-symplectic framework). Let $Z(x,t) = \widehat{Z}(kx + \omega t)$. Then $\widehat{Z}$ satisfies the ODE

$$\omega \mathbf{M}\widehat{Z}_\theta + k\mathbf{K}\widehat{Z}_\theta = \nabla S(\widehat{Z}), \qquad \theta = kx + \omega t. \tag{4.3}$$

However let

$$A(\widehat{Z}) = \frac{1}{2\pi} \int_0^{2\pi} \tfrac{1}{2}\langle \mathbf{M}\widehat{Z}_\theta, \widehat{Z}\rangle \, d\theta, \qquad B(\widehat{Z}) = \frac{1}{2\pi} \int_0^{2\pi} \tfrac{1}{2}\langle \mathbf{K}\widehat{Z}_\theta, \widehat{Z}\rangle \, d\theta,$$

where $\langle \cdot, \cdot \rangle$ is the standard inner product on $\mathbf{R}^3$. Then $\nabla A(\widehat{Z}) = \mathbf{M}\widehat{Z}_\theta$ and $\nabla B(\widehat{Z}) = \mathbf{K}\widehat{Z}_\theta$ (with respect to an inner product that includes integration over x) and so (4.3) becomes

$$\omega \, \nabla A(\widehat{Z}) + k \, \nabla B(\widehat{Z}) = \nabla S(\widehat{Z}). \tag{4.4}$$

The form of the equations (4.4) leads to an interesting variational characterisation of periodic travelling waves as critical points of the functional S with the functionals A and B as constraints and then ω and k are Lagrange multipliers.

A standard result from Lagrange multiplier theory is that ω and k can be parametrised by I_1 and I_2, the values of the level sets of the constraints A and B, and such states are non-degenerate if

$$\Delta_k = \det \begin{bmatrix} \frac{\partial \omega}{\partial I_1} & \frac{\partial \omega}{\partial I_2} \\ \frac{\partial k}{\partial I_1} & \frac{\partial k}{\partial I_2} \end{bmatrix}$$

is nonzero. However one can prove the following result.

Theorem 2 *If $\Delta_k > 0$, then the periodic travelling wave is linearly unstable.*

Proof: This result is proved in a similar way to Theorem 1. Let $Z = \widehat{Z}(\theta) + U(\theta, t)$ and then U satisfies

$$\mathbf{M}U_t + (k\mathbf{K} + \omega\mathbf{M})U_\theta = D^2 S(\widehat{Z})U, \tag{4.5a}$$

where $\widehat{Z}$ is 2π-periodic in θ. But, by Floquet's theorem, every bounded solution of (4.5a) is of the form $U(x,t) = e^{(\lambda t + i\alpha\theta)}\mathcal{V}(\theta)$ with $\mathcal{V}(\theta)$ 2π-periodic and satisfying

$$\mathbf{J}\mathcal{V}_\theta = (D^2 S(\widehat{Z}) - \lambda\mathbf{M} - i\alpha\mathbf{J})\mathcal{V} \qquad \text{with} \quad \mathbf{J} = k\mathbf{K} + \omega\mathbf{M}, \tag{4.5b}$$

and one can then apply the Lyapunov-Schmidt method as in Section 2 noting that the tangent vector to the periodic orbit satisfies (4.5b) when $\lambda = \alpha = 0$

and that there is a two parameter (I_1, I_2) family of such states. The full details, including the case of 3D waves, can be found in Bridges [4, 6].   ∎

The above theory also provides a framework for proving the instabilities predicted by the Whitham modulation theory. Here we show how the condition $\Delta_k > 0$ agrees precisely with the criterion predicted using the Whitham modulation theory (c.f. Whitham [16]). The model problem (4.1) is the Euler-Lagrange equation corresponding to the Lagrangian density

$$\widehat{\mathcal{L}} = \tfrac{1}{2}u_t^2 - \tfrac{1}{2}u_x^2 - V(u). \tag{4.6}$$

In the Whitham theory the Lagrangian density $\widehat{\mathcal{L}}$ is averaged to obtain

$$\mathcal{L} = \int_0^{2\pi} [\tfrac{1}{2}\omega^2 u_\theta^2 - \tfrac{1}{2}k^2 u_\theta^2 - V(u)]\, d\theta,$$

where $\theta = kx - \omega t$. Then $\mathcal{L}$, ω and k are taken to be slowly varying functions of x and t resulting in the Whitham modulation equations

$$\frac{\partial A}{\partial t} - \frac{\partial B}{\partial x} = 0,$$

$$\frac{\partial k}{\partial t} + \frac{\partial \omega}{\partial x} = 0,$$

where $A = \mathcal{L}_\omega$ and $B = \mathcal{L}_k$ or, letting A and B depend on ω and k,

$$\begin{bmatrix} A_\omega & A_k \\ 0 & 1 \end{bmatrix} \begin{bmatrix} \omega \\ k \end{bmatrix}_t + \begin{bmatrix} -B_\omega & -B_k \\ 1 & 0 \end{bmatrix} \begin{bmatrix} \omega \\ k \end{bmatrix}_x = 0. \tag{4.7}$$

Let

$$\begin{bmatrix} \omega \\ k \end{bmatrix} = \begin{bmatrix} \omega_0 \\ k_0 \end{bmatrix} + \begin{bmatrix} \widehat{\omega} \\ \widehat{k} \end{bmatrix} e^{i(\alpha x - \Omega t)},$$

where (ω_0, k_0) represents a basic steady travelling wave. Then substitution into (4.7) results in

$$\Omega = \frac{1}{A_\omega}(-A_k \pm \sqrt{-\Delta_{\mathcal{L}}}) \quad \text{where} \quad \Delta_{\mathcal{L}} = \det \begin{bmatrix} A_\omega & A_k \\ B_\omega & B_k \end{bmatrix}.$$

In other words if $\Delta_{\mathcal{L}} > 0$ there exists an unstable solution of the Whitham modulation equations. To see that this agrees with the criterion $\Delta_k > 0$, identify A and B with the functionals A and B in (4.4). Then noting that $\Delta_{\mathcal{L}} = 1/\Delta_k$ the correspondence follows. The advantage of the framework presented at the beginning of this section is two-fold: one works directly with the governing equations, without recourse to approximate modulation equations and secondly the theory is rigorous. Further results on proving instability results predicted by the Whitham theory as well as instability criteria for more general classes of waves are given in Bridges [4, 5, 6].

5 Further Results on Instability of Waves

In this section we sketch some other recent results on wave dynamics and instability.

(a) **Benjamin-Feir instability.** One of the most important instabilities in the theory of water waves is the Benjamin-Feir instability where a periodic Stokes' travelling wave loses stability to sideband perturbations (Benjamin & Feir [2]). An analysis of the long-time existence and well-posedness of the equations governing water waves or a direct analysis of the spectral problem for the time-dependent equations will be difficult and in some cases intractable. However to prove the existence of a subset of unstable eigenvalues of the linear stability problem for the Stokes' travelling wave is tractable and is sufficient to prove instability. In Bridges & Mielke [8], a rigorous proof of the Benjamin-Feir instability is given. The proof uses a sequence of centre-manifold reductions on the governing equations and the linear stability problem, reformulated as spatial evolution equations.

(b) **Pattern formation on the open ocean.** In the examples presented so far the periodic or quasiperiodic patterns were aligned with one space dimension. However the framework in Section 4 extends to the analysis of instability of multi-dimensional periodic or quasiperiodic patterns. A problem of particular interest is pattern formation on the open ocean: what are all possible periodic (quasiperiodic, etc.) patterns that can exist on the ocean surface and which are stable? This question is further complicated for oceanic flows, when the depth of the fluid is shallow, by the presence of mean flow (for example currents generated by the creation of patterns). While a complete characterisation of the linear stability problem is intractable in this case, significant progress can be made on the *instability* problem; a generalised class of sideband perturbations (sidebands in each spatial direction) *including mean flow effects* can be formulated and leads to a subclassification of unstable multi-periodic patterns (c.f. Bridges [5]).

(c) **Waves with weak dissipation.** Travelling waves can be characterised as relative equilibria and in conservative systems relative equilibria have an interesting geometric characterisation. A natural question that arises when analysing conservative model equations, such as the KdV equation or nonlinear wave equations, is the effect of weak dissipation. Recent results of Derks, Lewis & Ratiu [10] (and references therein) present a geometric framework for analysing the effect of weak dissipation on relative equilibria.

(d) **Instability of solitary waves.** The instability problem for solitary waves has a rich history. We will mention here some recent work where,

for a particular class of conservative systems, geometric properties of
the basic state are related to unstable eigenvalues. In the recent work
of Pego & Weinstein [15] (which also contains many references to previous work), a class of models is considered where the solitary wave can
be characterised as a critical point of the energy on level sets of the
momentum I with c, the speed of the wave, as a Lagrange multiplier.
Under suitable hypotheses, if the slope in the (I, c) plane is negative,
for a given solitary wave, it is linearly unstable. The proof of this result uses the Evans function formulation for eigenvalues of the linear
stability problem (c.f. Pego & Weinstein [15] for details and references).

(e) **Absolute and convective instabilities.** It has been known since the
early 1960's that unstable waves in homogeneous media can evolve in
two distinct ways. If a wave is exponentially growing in time at every
fixed $x_0 \in \mathbf{R}$ then we say that the wave is *absolutely unstable*. However
if, relative to a fixed frame, the wave is exponentially decaying but,
relative to a suitable *moving frame*, it is exponentially growing in time,
then we say that it is *convectively unstable*. This distinction is of great
importance in applications; if a wave is convectively unstable it may
convect away from a region of interest and may be controllable whereas
an absolutely unstable wave is growing at all points in space and therefore leads to catastrophic changes in the flowfield. Recently Brevdo &
Bridges [3] have extended this classification to the instability of spatially periodic travelling waves and flows. The theory is used to prove
that the Eckhaus instability (Eckhaus [11]) is an absolute instability
whereas periodic travelling wave solutions of the complex Ginzburg-
Landau equation have both convective and absolute instabilities with
a well-defined transition between the two states which cannot be predicted by studying the spectrum alone. The classification of secondary
instabilities in the Navier-Stokes equations is also formulated.

References

[1] V.I. Arnold. *Mathematical Methods of Classical Mechanics*, Springer-
Verlag, Heidelberg, 1978.

[2] T.B. Benjamin and J. Feir. The disintegration of wavetrains on deep
water. Part 1. Theory, *J. Fluid Mech.* **27**, 417-430, 1967.

[3] L. Brevdo and T.J. Bridges. Absolute and convective instabilities of spatially periodic flows, *Phil. Trans. Roy. Soc. Lond. A*, to appear, 1996.

[4] T.J. Bridges. Hamiltonian structure of plane-wave instabilities, *Field
Inst. Comm.* **8**, 19-33, 1996.

[5] T.J. Bridges. Periodic patterns, linear instability, symplectic structure and mean-flow dynamics for 3D surface waves, *Phil. Trans. Roy. Soc. Lond. A* **451**, in press, 1996.

[6] T.J. Bridges. Multi-symplectic structures and wave propagation, *Math. Proc. Camb. Phil. Soc.* **121**, in press, 1996.

[7] T.J. Bridges and A.J. Cooper. Spanwise modulation of streamwise rolls in rotating channel flow, *Quart. J. Mech. Appl. Math.* **48**, 257-284, 1995.

[8] T.J. Bridges and A. Mielke. A proof of the Benjamin-Feir instability, *Arch. Rat. Mech. Anal.* **133**, 145-198, 1995.

[9] T.J. Bridges and G. Rowlands. Instability of spatially quasiperiodic states of the Ginzburg-Landau equation, *Proc. Roy. Soc. Lond. A* **444**, 347-362, 1994.

[10] G. Derks, D. Lewis and T. Ratiu. Approximations with curves of relative equilibria in Hamiltonian systems with dissipation, *Nonlinearity* **8**, 1087-1114, 1995.

[11] W. Eckhaus. *Studies in Nonlinear Stability Theory*, Springer-Verlag, Heidelberg, 1965.

[12] E. van Groesen and E.M. de Jager. *Mathematical Structures in Continuous Dynamical Systems*, North-Holland, Elsevier, Amsterdam, 1994.

[13] E. Infeld and G. Rowlands. *Nonlinear Waves, Solitons and Chaos*, CUP, Cambridge, 1990.

[14] W. Magnus and S. Winkler. *Hill's Equation*, Dover Publ., New York, 1979.

[15] R.L. Pego and M.I. Weinstein. Eigenvalues and instabilities of solitary waves, *Phil. Trans. Roy. Soc. Lond. A* **340**, 47-94, 1992.

[16] G.B. Whitham. *Linear and Nonlinear Waves*, Wiley-Interscience, New York, 1974.

Non-Smooth Dynamical Systems and the Grazing Bifurcation

C.J. Budd
School of Mathematics,
University of Bath,
Bath

1 Introduction

Many classical treatments of dynamical systems make the underlying assumption that the systems they are describing are smooth. Whilst this assumption is reasonable for most physical problems there are also systems which are best described (at least on macroscopic scales) in terms of non-smooth dynamics. The simplest such example (an an early introduction to dynamical systems for many of us) is the trajectory observed when an elastic ball is dropped on to a hard floor. Many other similar examples exist in mechanical and electrical systems.

At first it would appear that a systematic study of such systems is hopeless, after all, the study of smooth dynamical systems is hard enough! However, by making reasonable assumptions about the departure of the system from smoothness, some generic properties emerge and we are able to identify a new form of bifurcation – the *grazing bifurcation*. This bifurcation arises when, under a change of parameters, the system evolves so that a smooth motion first starts to lose regularity. An example of this occurs when a freely oscillating cantilever beam is brought closer to an obstacle. At some point an impact will occur and, remarkably, this simple experiment often leads to complex chaotic motion. The grazing bifurcation describes a mechanism for this transition and may (possibly) account for many other examples of chaotic vibration observed in mechanical systems. The grazing bifurcation has now been observed in many non-smooth systems, and part of the purpose of this article is to motivate it and to describe some of the recent exciting mathematical results concerning it. We will conclude with some open problems.

2 Examples of Non-Smooth Dynamical Systems

We start our discussion by considering some systems which arise naturally and which have non-smooth dynamics. Such problems arise in many fields but were perhaps first considered by Fermi [9], who was considering the origin of cosmic rays. He proposed that they might originate from the exchange of charged particles between large gaseous clouds. A greatly simplified model of this comprises a ball in a cylinder accelerated by a moving piston. The ball typically moves to the end wall of the cylinder, impacts, and then rebounds, hitting the cylinder again and thus continuing the motion. Thus the motion of the ball is a sequence of smooth motions with instantaneous changes in velocity. This system is Hamiltonian and has received much attention in the theoretical physics literature where it is shown, for example, that a periodically moving cylinder can lead to chaotic motion of the ball. A similar model has been used by True [34] and Knudsen, Feldberg & True [17] to describe the motion of railway bogies. Closely related is the problem of finding the motion of an elastic ball moving under gravity and dropped on to a periodically moving cylinder. This system was introduced by Pustylnikov [29] and described in detail by Guckenheimer & Holmes [14]. Again, a periodic excitation can lead to chaotic motion of the ball. A demonstration of this behaviour has been constructed by Popp [26] and it is also used as an example of controlling chaotic systems.

This system can be extended by considering the ball to be suspended on a spring. If it is struck from below by a moving obstacle then we observe a combination of the free motion of the ball on the spring together with the forced motion due to the impacts. Similarly, if the ball is suspended above a fixed obstacle and oscillated from above, then small motions on the spring will not bring it into contact with the obstacle and hence will be periodic. Larger motions will however result in impacts and it is this transition which is of interest to us. Typically we observe chaotic motion or large amplitude periodic motion with high velocity impacts between the ball and the obstacle. Very closely related are problems concerning harmonic oscillators with springs that have stiffness that depends discontinuously upon position, for example the motion of the system described by

$$\frac{d^2x}{dt^2} + K(x)x = \cos(\omega t), \qquad (2.1)$$

where

$$K(x) = \begin{cases} k_1 & \text{if } x < x_0, \\ k_2 & \text{if } x > x_0. \end{cases} \qquad (2.2)$$

These systems, which have been used by Thompson & Stewart [33] to study the motion of ships in a harbour and by Doole & Hogan [8] to look at the motion of suspension bridges, are analysed in the seminal paper by Shaw & Holmes [31]. If k_2/k_1 is large and k_1 is not small then they have very similar dynamics to the simple impacting model. Problems with $k_2 = 0$ occur in systems with play or backlash between components (see Kleczka, Kreuzer & Schiehlen [16]). Both the ball on a spring and the piecewise linear models above are very closely related to the motion of an elastic beam impacting an obstacle for which some experimental results are presented in Cusumano, Sharkady & Kimble [7] (see also [32]), or of tubes (in say a heat exchanger) excited by a flow and vibrating against their supports [13]. These systems are more complex, however, as the beam or tube may vibrate in many modes. Another very natural physical system giving rise to discontinuous dynamics is the stick-slip behaviour of systems with friction [27, 28] including the motion of turbine blades [25]. In such systems the friction force is always in the direction opposite to that of the motion. If we consider a forced mass on a spring which is in friction contact with a moving obstacle, then a simplified model of the motion is given by

$$\frac{d^2x}{dt^2} + \mu \operatorname{sgn}\left(\frac{dx}{dt}\right) + x = \phi(t), \tag{2.3}$$

where $\operatorname{sgn}(x) = 1$ if $x > 0$ and $\operatorname{sgn}(x) = -1$ if $x < 0$. If $\phi(t) = \sin(t)$ then such an oscillator can exhibit periodic or chaotic motion. (The periodic motion is well known as, for example, wheel squeal in railway bogies or in the squeaking of chalk on a black-board.) Again, grazing phenomena are observed when a system just starts to slip as the friction force becomes too weak to maintain equilibrium. Behaviour very similar to grazing also appears to occur in electrical circuits. Indeed the simulations of DC-DC converters described by Hamill, Deane & Jefferies [15] are qualitatively very similar to the results obtained from a grazing bifurcation.

As in any dynamical system it is natural to ask questions such as when the system will be in equilibrium, when it will have periodic states, when it will be chaotic and what the transitions will be between the states. We now study these questions by looking at the particular example of the impact oscillator.

3 The Impact Oscillator and its Associated Maps

The impact oscillator is the idealisation of the problem comprising the ball on a spring hitting an obstacle. This system has been studied in the Russian and Czech literature [24] for some time but has more recently aroused a great deal of interest, both for its (obvious) applications and the interesting nature of the mathematics describing it. The simplest impact oscillator is described by the system

$$\frac{d^2x}{dt^2} + \zeta \frac{dx}{dt} + x = \cos(\omega t), \quad x < d, \tag{3.1}$$

$$\frac{dx}{dt} \to -\lambda \frac{dx}{dt}, \quad x = d, \tag{3.2}$$

(where $\lambda < 1$ is the coefficient of restitution), so that it has a series of impacts with the obstacle located at $x = d$ at times t_i with velocities v_i. Between impacts its motion is smooth. An experimental realisation of this oscillator using gliders on an air-bed is described in [30].

The impact oscillator may exhibit both periodic and chaotic behaviour depending upon the parameters λ, d, ω and ζ with transitions between these states described by both smooth and grazing bifurcations. The simplest such motion is periodic with period

$$T = \frac{2\pi n}{\omega}, \quad n = 1, 2, 3, \ldots .$$

Such a motion typically repeats after m impacts such that

$$v_{i+m} = v_i \quad \text{and} \quad t_{i+m} = t_i + T$$

and is referred to as an (m, n) or an m/n orbit. Of these orbits the $(1, n)$ are the most common, are easiest to analyse, and typically have the largest amplitude and domains of attraction. It is straightforward to write down algebraic conditions for these orbits [31, 35, 36, 3] and to analyse their (linear) stability under small perturbations. In many respects they demonstrate the same features as fixed points of a smooth dynamical system in that they have attracting manifolds and may lose stability through period-doubling or saddle-node bifurcations. However, an alternative mechanism for the loss of stability arises when an additional impact occurs at a time t such that $t_i < t < t_i + T$. If this first occurs at a given parameter value, say $d = d_0$ (so that the intermediate impact has zero velocity), then the periodic orbit encounters a *grazing bifurcation* at d_0 which typically leads to chaotic or more complex behaviour.

A convenient way to study the impact oscillator is by considering a map P relating one impact to the next. It can be shown that if an impact occurs with (positive) velocity v_i at a time t_i then the velocity and time (v_{i+1}, t_{i+1}) of the next impact are well defined functions of v_i and the phase $\phi_i = (t_i)_{\text{mod}(T)}$. This leads us to consider the map

$$P : (\phi_i, v_i) \rightarrow (\phi_{i+1}, v_{i+1}).$$

The structure of this map is described in detail in [3]. It can be constructed easily numerically and is a natural way to study the dynamics of the impact oscillator. It maps the cylinder to itself and an (m, n) periodic orbit is a fixed point of the iterated map P^m. Moreover, P is smooth apart from on a one-dimensional (branched) set S defined by

$$S = \{(\phi, v) : P(\phi, v) = (\psi, 0)\} .$$

Across this set P is discontinuous and introduces infinite local stretching into the phase space. To consider the action of this map, let S be the set defined above and W the image set $P^2(S)$. The nature of the discontinuity of P is that points on one side of S and close to S are mapped by P to points close to W and on one side of W. In contrast points on the other side of S are first mapped to a neighbourhood of the line $v = 0$ and then are mapped to the other side of W on the second iterate of P. More significant is the stretching that this map introduces. To show this we introduce a new (local) coordinate system (x, y) so that $y = 0$ for a point on S and x and y represent coordinates in directions tangent to and orthogonal to S respectively. Here we consider the points for *positive* y to be first mapped to a neighbourhood of $v = 0$ and then to a neighbourhood of W. Similarly we introduce x' and y' which are respectively parallel and orthogonal to W. Then, in this new coordinate system, Nordmark [20] (see also [1]) showed that locally P has the following form:

$$\text{if} \quad y < 0 \quad \text{then} \quad (x', y') = P(x, y) = (ax + by, cy), \quad (3.3)$$

$$\text{if} \quad y > 0 \quad \text{then} \quad (x', y') = P^2(x, y) = (ax + dy, e\sqrt{y}), \quad (3.4)$$

where the values of a, b, c, d and e are chosen suitably.

The most important feature of this map is that it has a square-root singularity as $y \to 0$ which leads to infinite local stretching in this limit. As an immediate consequence of this stretching we see complex chaotic behaviour. For example, if a set K of data intersects the set S then the action of P is to stretch and fold K leading to an action on K closely resembling that of a horseshoe map. Furthermore, if a fixed point of P changes under the variation of a parameter so that it intersects S then the action of the stretching destabilises this point, leading to the grazing bifurcation and complex motion.

We now give two examples of the motion at some specific parameter values which demonstrate some of the dynamics possible.

3.1 Resonant motion, period doubling and grazing

If $d = 0$, $\lambda = 0.8$ and $\zeta = 0$, then a unique $(1, 1)$ orbit is stable in a neighbourhood of $\omega = 2$ and has a large (resonant) amplitude and impact velocity v_1. For certain (narrowly defined) parameter values, this orbit co-exists with a $(3, 3)$ or with a (chattering) $(\infty, 6)$ orbit [2]. Due to the stretching associated with the map P, the domains of attraction of these more complex orbits can be large, complex and tangled with that of the $(1, 1)$ orbit. In a typical experiment, where ω is increased steadily and the behaviour of the system observed, these coexisting orbits might appear to be a thin band of noise superimposed on the simpler $(1, 1)$ orbit. Possible behaviour of this nature has been observed by Hamill, Deane & Jefferies [15]. As ω increases (or decreases) the simple $(1, 1)$ orbit loses stability at a smooth super-critical period-doubling bifurcation to a $(2, 2)$ orbit with impact velocities v_1 and v_2. The $(2, 2)$ orbit period doubles again as ω increases to a $(4, 4)$ orbit with impact velocities v_1, v_2, v_3 and v_4. Chaotic behaviour with an associated strange attractor is observed when $\omega = 2.8$. As ω is increased further, the

$(2, 2)$ orbit restabilises and is globally attracting in a range of ω centred on $\omega = 3$. The orbit is then destroyed at a grazing bifurcation at $\omega = 3.01$ when an intervening impact occurs between the existing two impacts. There then follows a period of chaotic motion before a simple globally attracting $(1, 2)$ orbit arises at $\omega = 3.35$. This behaviour is presented in Fig. 1 which shows the values of the impact velocity v_i that occur after 1500 impacts with initial values of $\phi = 0.5$, $v = -1.5$. A detailed analysis of the transition from the $(1, 1)$ to $(2, 2)$ orbit and the consequent restabilisation of the $(2, 2)$ orbit is given in [4].

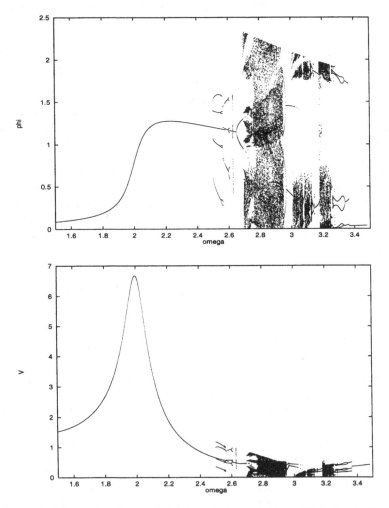

Figure 1: Bifurcation diagram showing the impact phase (ϕ) and velocity (v) with $d = 0$, $\lambda = 0.8$, $\phi_0 = 0.5$, $v_0 = -1.5$ and 3000 initial transients.

3.2   First impacting phenomena

If $d > 1$ and we take the same parameter values as above, then for small values of ω the forced motion of the oscillator will be too small for any impacting to occur. However, as ω is increased toward the natural frequency of the oscillator, then the amplitude of the forced motion increases and there is a first value of $\omega = \omega_1$ given by the smaller root of the equation

$$1/d = \sqrt{(\zeta\omega)^2 + (\omega^2 - 1)^2} \tag{3.5}$$

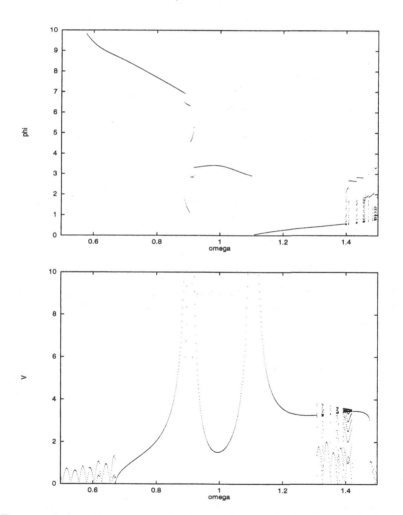

Figure 2: Bifurcation diagram showing the impact phase (ϕ) and velocity (v) with $d = 2$, $\lambda = 0.8$, $\phi_0 = 0.5$, $v_0 = -1.5$ and 3000 initial transients.

(so that $\omega_1 \approx \sqrt{1 - 1/d}$ when ζ is small), when an impact of zero velocity first occurs. For subsequent values of ω the motion is most complex and the transition from the simple non-impacting motion to this complex motion is described by the grazing bifurcation [10, 11, 12]. Interestingly, for larger values of ω, the motion stabilises to a simple $(1, 1)$ periodic orbit which is globally attracting, but which also has regions of parameter values for which it coexists with other more complex orbits. There is then a subsequent value of $\omega = \omega_{SN} \approx 1.45$ where the $(1, 1)$ orbit experiences a simple saddle-node bifurcation and loses stability. In this case the oscillator returns to a non-impacting state. If ω is now *reduced* then there is a further value of ω given by the larger root of (3.5), so that if ζ is small then

$$\omega_2 \approx \sqrt{1 + 1/d}, \quad \text{where} \quad \omega_1 < 1 < \omega_2 < \omega_{SN},$$

when a zero velocity impact takes place and the motion of the oscillator is again complex. There is thus a natural hysteresis in the behaviour of the impact oscillator as a function of ω with complex motion generated by grazing.

We illustrate the corresponding dynamics when $d = 2$, $\lambda = 0.8$ and $\zeta = 0.1$ in Fig. 2. This figure also includes other orbits (particularly in the case $\omega < \omega_1$) where other types of impacting motion occur.

Other grazing bifurcations can be observed if ω is fixed and d varies. For example, if $\omega = 2$ and d is reduced from zero, then a grazing bifurcation is observed close to $d = -1/3$. A description of this is given in [1].

4   Grazing and the Grazing Bifurcation

The examples described above naturally lead to the important question as to what behaviour is observed when the system parameters vary from a periodic state with no impacts (or high velocity impacts) to one where an impact first occurs (or there is an additional low velocity impact between those of high velocity)? Similar questions arise in studies of the bouncing ball or in friction oscillators. As remarked in Section 3, this corresponds to a fixed point (or orbit) of the map P intersecting the set S as a parameter is varied. The stretching of the phase space close to S gives rise to new bifurcation phenomena which appear to be universal to many systems. The corresponding bifurcation is called a *grazing bifurcation* as it is associated with a low velocity *grazing* impact. This bifurcation was first observed by Peterka [24] and discovered independently by Whiston [35, 36, 37] and Nordmark [20, 21, 22]. Subsequently it has been investigated analytically by Foale & Bishop [11, 12], Budd & Dux [1], Lamba & Budd [18] and in a series of papers by the group at Maryland [23, 5, 6]. Some experimental results have also been reported in [11, 12] and in the recent work by Mullin [19]. In this presentation we follow that of [22, 23] quite closely.

Suppose now that for some parameter d (which may for example be the clearance of the system) a $(1, n)$ orbit exists and is linearly stable and that this orbit has a local maximum between impacts (or suppose similarly that an orbit of small amplitude without impact exists). As d is changed, there will typically be a first value (without loss of generality at $d = 0$) at which a grazing impact first occurs. Typically such an impact introduces a destabilising effect on the $(1, n)$ orbit due to the large stretching of the phase space and a different motion results. If the $(1, n)$ orbit is stable for $d < 0$ and unstable for $d > 0$ then such a bifurcation is *supercritical* with the resulting motion confined to a neighbourhood of $\sqrt{d}$ of the original orbit [20]. Depending upon other parameters in the system, four types of behaviour are then observed.

(i) <u>Large coefficient of restitution</u>: There is an immediate creation of an intermittent chaotic motion together with a large set of unstable periodic orbits. There is a non-zero interval $0 < d < d'$ in which there are no periodic windows. The chaotic motion lies on an attracting set of size proportional to $\sqrt{d}$. It is shown in [18] and [1] that the number of unstable periodic orbits and the reciprocal of the Lyapunov exponents both scale as $\log(d)$.

(ii) <u>Intermediate coefficient of restitution</u>: There are an infinite sequence of periodic windows alternating with chaotic bands without windows. The windows are period *adding* so that if there is a window at some value of d containing an n-impact orbit, then there is a next window (as d decreases) containing an $(n+1)$-impact orbit. The width (and location) of the windows decreases geometrically as $d \to 0$.

(iii) <u>Low coefficient of restitution</u>: No chaotic motion exists, but the period-adding windows do exist and overlap giving multiple periodic states and an infinite sequence of periodic motions.

(iv) The $(1, n)$ orbit and an unstable orbit coalesce at $d = 0$.

An example of a bifurcation taking the form of case (ii) is presented in Fig. 3 and of case (iii) in Fig. 4.

It is most interesting to compare this behaviour with that of the bifurcation diagram typical of the logistic map and described by Sarkovski's Theorem. In this case there are no intervals of purely chaotic motion which do not contain periodic windows. Furthermore the relation between the periodic orbits in the windows is somewhat more complex than period-adding. Indeed, it is period-adding which is the strongest indicator of a grazing bifurcation.

We now show briefly why this behaviour occurs. To do this we follow the elegant treatment given in [22] and consider the bifurcation associated with a one-dimensional map which inherits many of the features of the impact map. A similar approach is taken in [11, 12]. In [1, 23] these arguments are

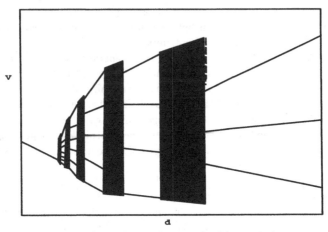

Figure 3: A grazing bifurcation with period adding windows separated by bands of chaotic behaviour.

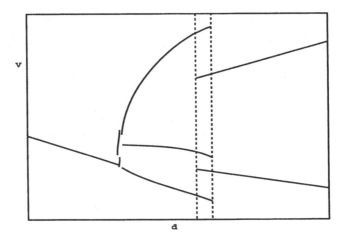

Figure 4: A grazing bifurcation with coincident period adding windows.

generalised to the true two-dimensional map. However, this generalisation is relatively straightforward as, close to the bifurcation, the important action of the map P is largely confined to the one-dimensional stable manifold of the fixed point (where we note that the manifold persists even if the fixed point itself is destroyed in the graze).

Accordingly, motivated by the local form of the map P close to S given in

(3.3) and (3.4), we look at the dynamics of the map $f(x)$ defined in [22] by

$$f(x) = \begin{cases} \sqrt{d-x} + \lambda x, & x \leq d, \\ \lambda x, & x \geq d, \end{cases} \qquad (4.1)$$

where $\lambda < 1$ is equivalent to the coefficient of restitution. The most important feature of this map is the square root singularity which introduces locally unbounded stretching. For this map, d plays the role of a bifurcation parameter.

If $d < 0$ then $f(x)$ has a single (stable) fixed point at $x = 0$. This loses stability if $d > 0$ and for small $d > 0$ we observe motion consistent with the first three of the cases above, with case (i) if $2/3 < \lambda < 1$, case (ii) if $1/4 < \lambda < 2/3$ and case (iii) if $0 < \lambda < 1/4$.

We now show how this behaviour arises when $d > 0$. It can be shown easily, that if $\lambda d < x_0 < d$ and if d is sufficiently small, then due to the stretching associated with the square root, the first iterate $x_1 = f(x_0)$ of f will lie in the region $x_1 > d$ and that there will subsequently be a (possibly large) number $m(x_0, d)$ of iterates x_n of $f(x)$ with $x_n > d$ if $n \leq m$, such that x_{m+1} reenters the interval $I \equiv \{x : \lambda d < x < d\}$. Thus the set I acts as a trapping region for the flow. The value of $m(x_0, d)$ takes its maximum value $M(d)$ when $x_0 = \lambda$ and decreases monotonically as x_0 increases so that it is constant on subintervals of I. In a typical motion an iteration lies in I once and in the range $x > d$ more frequently. A periodic orbit with this property is referred to as a *maximal orbit* in [23].

Motivated by this discussion we introduce a map $F : I \to I$ from the interval $[\lambda d, d]$ to itself defined by

$$F(x_0) = x_{m+1},$$

which is continuous on those subintervals for which m is constant, and is discontinuous across the interval boundaries. Thus F has M continuous branches, and on each such branch it takes the form

$$F(z) = \frac{\lambda^m}{\sqrt{d}}\sqrt{1-z} + \lambda^{m+1}z,$$

where $z = x/d$. (See [1] for a very similar description of a corresponding two-dimensional map.) It is shown in [20] (and can be verified easily) that decreasing d by a factor of λ^2 increases M by one and adds an extra branch to F but otherwise the form (and magnitude) of F does not change. To formalise this Nordmark [22] introduced a new parameter $\mu \in (\lambda^2, 1]$ defined by

$$\mu \equiv \left(\frac{\lambda}{F(\lambda)}\right)^2$$

and then for large M,

$$\mu \approx \frac{d}{\lambda^{2(M-1)}(1-\lambda)}. \tag{4.2}$$

Under this change of variables, and putting $k = M - m$, we get

$$F(z) = \frac{\lambda\sqrt{1-\lambda}}{\sqrt{1-\lambda}\sqrt{\mu}\lambda^k} + \frac{\lambda^{M+1}}{\lambda^k}\left(z - \lambda\frac{\sqrt{1-z}}{\sqrt{1-\lambda}}\right) \approx G(z)$$

where

$$G(z) = \frac{\lambda\sqrt{1-z}}{\sqrt{1-\lambda}\sqrt{\mu}\lambda^k}. \tag{4.3}$$

The map $G(z)$ is illustrated in Fig. 5.

We see from this result that if μ is constant, then F is well approximated by the universal function G. Now from (4.2) it is evident that reducing d by a factor of λ^2 will leave μ the same provided that M is increased by one. The dynamics of the map F will be essentially the same in these two cases and will be described by the dynamics of the map G although M is increased by one. This is the mechanism behind the period-adding phenomena.

The map G is continuous on the intervals I_k defined by

$$I_0 = [\lambda, B_0], \quad I_k = (B_{k-1}, B_k], \quad k = 1, 2, 3, \ldots$$

where

$$B_k = 1 - \mu\lambda^{2k}(1-\lambda).$$

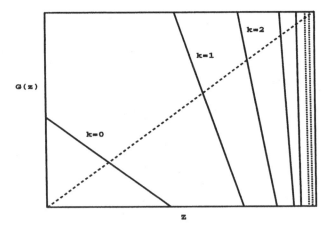

Figure 5: The map $G(z)$ showing the disconnected branches.

The map G is essentially independent of the function f, provided that this function has a square root singularity, and gives a *normal form* for the grazing bifurcation which extends to higher dimensions. We now look at the dynamics of its iterates.

A simple calculation shows that

$$G'(z) < G'(\lambda) = -\frac{\lambda}{2(1-\lambda)\sqrt{\mu}} \qquad (4.4)$$

and that G has an infinite number of fixed points z_k with

$$G'(z_k) = -\frac{z_k}{2(1-z_k)}.$$

As $\mu \in (\lambda^2, 1]$ it follows from (4.4) that if $\lambda > 2/3$ then $G'(z) < -1$ so that no periodic orbits exist and all motion is chaotic.

For each fixed M, if $\lambda < 2/3$ then it is possible for z_0 to be stable. In particular it exists if $\lambda < 2/3$ and if μ is close to one. For the map f such a fixed point corresponds to a so called *maximal M-orbit* which has one iterate in the region where the map takes square root form and M iterates in the region where the map is linear. This orbit will then exist if $d < d_n$, where d_n is obtained by setting $\mu = 1$ in (4.2), so that

$$d_n(M) \approx (1-\lambda)\lambda^{2(M-1)}.$$

For $d > d_n$ the maximal M-orbit does not exist and it vanishes discontinuously as d increases. Now if $1/4 < \lambda < 2/3$, it follows from (4.4) that the periodic orbit has a super-critical period doubling bifurcation when $\mu = \mu_f$ given by

$$\mu_f = \frac{3\lambda^2}{4(1-\lambda)},$$

with the corresponding value of

$$d_f(M) \approx \frac{3}{4}\lambda^{2M}.$$

Thus, for each fixed M we will see an M-maximal orbit in the interval $d_f(M) < d < d_n(M)$ and we note that

$$\frac{d_f(M+1)}{d_f(M)} \approx \frac{d_n(M+1)}{d_n(M)} \approx \lambda^2.$$

If $1/4 < \lambda < 2/3$, then Nordmark [22] shows that no periodic orbits of $G(z)$, $G(G(z))$ or of higher iterates exist and consequently the dynamics is completely chaotic in this interval. This picture then explains the existence of period adding windows separated by intervals of chaos.

We now finally look at the case of $\lambda < 1/4$. In this case there are no period doubling bifurcations and the M-orbit persists as d is reduced. Furthermore, the second fixed point z_1 of the map G becomes stable so that we see a coincidence of an M-orbit and an $(M+1)$-orbit. Thus in this case we no longer have periodic windows separated by chaotic bands, but instead see only periodic behaviour as d is reduced to zero.

5 Conclusions

We have shown that complex dynamic behaviour can be found at a grazing bifurcation and that a normal form of this bifurcation can be constructed to determine the resulting dynamics. These results lead to several interesting open questions. For example, will grazing bifurcations be observed in all of the systems discussed in Section 2? There are strong indications of very similar behaviour to grazing in the friction oscillators described in [27, 28] and in the DC-DC converters in [15] but much work needs to be done to see whether grazing is the fundamental cause of the instabilities observed. A useful study could also be made of grazing phenomena in control systems. Part of this investigation will involve the study of the robustness (and consequent generality) of the grazing bifurcation under changes in the system. As an example of this we could investigate the grazing phenomenon as a global limit of smooth bifurcations (for example if the rigid impact is replaced by a smooth one). Other open questions are concerned with a rigorous proof of the globally attracting nature of the periodic orbits in the period-adding windows and a more subtle study of the way that these lose stability. Much remains to be done!

6 Acknowledgements

I would like to thank Mr G. Lee and Mr J. Mulhern for their help in preparing the figures and Mr M. di Bernardo for some very useful conversations on the role of grazing in DC-DC converters.

References

[1] Budd, C. and Dux, F., Intermittent behaviour in an impact oscillator close to resonance, *Nonlinearity* **7**, 1191-1124, 1994.

[2] Budd, C. and Dux, F., Chattering and related behaviour in impact oscillators, *Phil. Trans. Roy. Soc. Lond. A* **347**, 365-389, 1994.

[3] Budd, C., Dux, F. and Cliffe, K.A., The effect of frequency and clearance variations on single-degree-of-freedom impact oscillators, *J. Sound Vib.* **184**, 475-502, 1995.

[4] Budd, C. and Lee, G., Double impact orbits of a single-degree-of-freedom impact oscillator subject to periodic forcing of odd frequency, submitted, 1995.

[5] Chin, W., Ott, E., Nusse, H. and Grebogi, C., Universal behaviour of impact oscillators near grazing incidence, Univ. Maryland Preprint, 1994.

[6] Chin, W., Ott, E., Nusse, H. and Grebogi, C., Grazing bifurcations in impact oscillators, Univ. Maryland Preprint, 1994.

[7] Cusumano, J., Sharkady, M. and Kimble, B., Experimental measurements of dimensionality and spatial coherence in the dynamics of a flexible beam impact oscillator, *Phil. Trans. Roy. Soc. Lond. A.* **347**, 421-438, 1994.

[8] Doole, S. and Hogan, S.J., The nonlinear dynamics of suspension bridges under harmonic forcing, to appear in *Dyn. Stab. Syst.*, 1995.

[9] Fermi, E., On the origin of cosmic radiation, *Phys. Rev.* **75**, 1169-1174, 1949.

[10] Foale, S., Analytical determination of bifurcations in an impact oscillator, *Phil. Trans. Roy. Soc. Lond. A* **347**, 353-364, 1994.

[11] Foale, S. and Bishop, S., Dynamical complexities of forced impacting systems, *Phil. Trans. Roy. Soc. A* **338**, 547-556, 1992.

[12] Foale, S. and Bishop, S., Bifurcations in impact oscillators: theory and experiments, *Proc. IUTAM Symp. Nonlin. Dyn. Chaos Engng Dynamics*, Wiley, 1994.

[13] Goyder, H. and Teh, C., A study of the impact dynamics of loosely supported heat exchanger tubes, *J. Pressure Vessel Technology* **111**, 394-401, 1989.

[14] Guckenheimer, J. and Holmes, P., *Nonlinear Oscillations, Dynamical Systems and Bifurcations of Vector Fields*, Springer-Verlag, 1983.

[15] Hamill, D., Deane, J. and Jefferies, D., Modelling of chaotic DC-DC converters using iterated nonlinear mappings, *IEEE Trans. Power Electronics* **7**, 25-36, 1992.

[16] Kleczka, M., Kreuzer, E. and Schiehlen, W., Local and global stability of a piecewise linear oscillator, *Phil. Trans. Roy. Soc. Lond. A* **338**, 533-546, 1992.

[17] Knudsen, C., Feldberg, R. and True, H., Bifurcations and chaos in a model of a rolling railway wheelset, *Phil. Trans. Roy. Soc. Lond.* **338**, 455-469, 1992.

[18] Lamba, H. and Budd, C., Scaling of Lyapunov exponents at non-smooth bifurcations, *Phys. Rev. E.* **50**, 84-91, 1994.

[19] Mullin, T., Experimental observations of vibro-impact in a cantilever beam, in preparation, 1995.

[20] Nordmark, A., Non-periodic motion caused by grazing incidence in an impact oscillator, *J. Sound Vib.* **145**, 279-297, 1991.

[21] Nordmark, A., Effects due to low velocity impact in mechanical oscillators, *Int. J. Bif. Chaos* **2**, 597-605, 1992.

[22] Nordmark, A., Non-smooth bifurcations in mappings with square-root singularities, preprint, 1993.

[23] Nusse, H., Ott, E. and Yorke, J., Border collision bifurcations: an explanation for observed bifurcation phenomena, *Phys. Rev. E* **49**, 1073-1076, 1994.

[24] Peterka, F., Transition to chaotic motion in mechanical systems with impacts, *J. Sound Vib.* **154**, 95-115, 1992.

[25] Pfeiffer, F. and Hajek, M., Stick-slip motions of turbine blade dampers, *Phil. Trans. Roy. Soc. Lond. A.* **338**, 503-517, 1992.

[26] Popp, K., Regelung eines technischen prozesses mit PEARL, Institut fur Regelungstechnik, Universitat Hannover Report, 1995.

[27] Popp, K. and Stelter, P., Nonlinear oscillations of structures induced by dry friction, in *Nonlinear Dynamics in Engineering Systems-IUTAM Symposium Stuttgart 1989*, ed. W. Schiehlen, Springer, 233-240, 1990.

[28] Popp, K. and Stelter, P., Stick-slip vibrations and chaos, *Phil. Trans. Roy. Soc. Lond. A.* **332**, 89-105, 1990.

[29] Pustylnikov, L., Stable and oscillating modes in nonautonomous dynamical systems, *Trans. Moscow Math. Soc.* **2**, 1-101, 1978.

[30] Reid, C. and Whineray, S., The resonant response of a simple harmonic half-oscillator, *Phys. Lett. A* **199**, 49-54, 1995.

[31] Shaw, S. and Holmes, P., A periodically forced piecewise linear oscillator, *J. Sound Vib.* **90**, 129-155, 1983.

[32] Stensson, A. and Nordmark, A., Experimental investigation of some consequences of low velocity impacts in the chaotic dynamics of a mechanical systems, *Phil. Trans. Roy. Soc. Lond. A* **347**, 439-448, 1994.

[33] Thompson, J.M.T. and Stewart, H.B., *Nonlinear Dynamics and Chaos*, Wiley, 1986.

[34] True, H., Chaotic motion of railway vehicles, *Proc. 11th IAVSD Symp. on Vehicle Systems*, ed. R. Anderson, Amsterdam/Lisse, Swets and Zeitlinger, 578-587, 1989.

[35] Whiston, G., The vibro-impact response of a harmonically excited and preloaded one-dimensional linear oscillator, *J. Sound Vib.* **115**, 303-324, 1987.

[36] Whiston, G., Global dynamics of a vibro-impacting linear oscillator, *J. Sound Vib.* **118**, 395-429, 1987.

[37] Whiston, G., Singularities in vibro-impact dynamics, *J. Sound Vib.* **152**, 427-460, 1992.

Solid Tumour Growth: A Case Study in Mathematical Biology

B.D. Sleeman
Department of Applied Mathematical Studies,
University of Leeds,
Leeds

1 Introduction

Tumours can arise from the cells of nearly all types of body tissue and this diversity of origin is largely responsible for the wide variety of the structural appearances of tumours. This diversity together with the complex and multifaceted nature of "in vivo" tumour growth is reflected in the many mathematical and theoretical models which have been proposed in an attempt to describe tumour growth and spread. These models draw on several mathematical disciplines ranging from applied analysis, involving ordinary and partial differential equations, as well as dynamical systems theory, stochastic models and computer simulations.

In this article we endeavour to describe the main features of tumour growth and in so doing, to draw attention to recent ideas and developments. In particular we focus on those areas where our understanding is limited and where mathematical modelling is likely to have a significant impact in the future. Included in this article is a fairly comprehensive bibliography which it is hoped will form a useful introduction to those wishing to enter this exciting and challenging field of applied mathematical research.

Before we can attempt to build models of tumour growth it is important to become acquainted with some of the basic notions and terminology.

2 Neoplastic Disease and the Growth of Tumours

Organ size and the number of cells it contains are usually maintained by the activity of control mechanisms that regulate the mitotic activity of cells. The control mechanisms are geared so as to

- permit repair of damaged or otherwise injured tissue;

- allow proliferation and the replacement of cells that undergo continuous wear and tear. Such is the case of the cells of the intestinal mucosa or of the skin;

- control hyperplasia of tissues in response to increased functional requirements. Hyperplasia is an excess of proliferation of cells occurring for example in the liver, salivary glands, pancreas, thyroid, adrenal cortex and ovaries etc.

Neoplasia is a state in which these control mechanisms become deficient and excessive proliferation of cells continues without relation to normal growth and tissue repair. This gives rise to a neoplasm, that is an abnormal tissue mass or tumour. On the basis of this discussion we may define a tumour as follows.

Definition 1 *A tumour is a mass of tissue formed as a result of abnormal, excessive and inappropriate proliferation of cells, the growth of which continues indefinitely and regardless of the mechanisms which control normal cellular proliferation.*

3   Classification of Tumours

Tumours are broadly speaking classified into one of two types: *benign* and *malignant*. To distinguish between these extremeties is often a difficult task, particularly at the early stages of growth. However a basic guiding principle is based on three factors. These are

(i) their degree of differentiation,

(ii) their rate of growth,

(iii) the manner in which they grow.

3.1   Differentiation

A tumour is highly differentiated when its structure bears a close resemblance to the tissue of origin. For example a well differentiated tumour of thyroid epithelium forms follicles, produces and stores thyroglobulin and may secrete thyroid-hormone. As well as retaining the functions of the original cells such tumours are readily distinguishable from non-neoplastic cells since they form a discrete mass.

At the other extreme, malignant tumours are often poorly differentiated in that they may have a premature undifferentiated appearance, producing a highly cellular mass in which little attempt at forming special structures or cell products can be discerned. Indeed cells of malignant tumours usually fail to differentiate fully and may vary abnormally in size and shape. This can be seen in certain tumours in which some of the cells are enormous and contain a single very large nucleus or multiple nuclei. Such cells have arisen by replication of the cellular DNA without complete mitosis.

3.2   Rate of growth

The rate of growth of tumour cells depends on the rate of cell production and the rate of cell loss. The rate of production of cells depends in turn on the number of cells undergoing mitosis and on the time they take to complete the cell cycle. It has been observed that malignant tumours in experimental animals show a logarithmic growth at first, but as the tumour enlarges, cell loss increases and the rate of growth tends to slow down. Thus, cells of malignant tumours, being abnormal, usually have a short mean life span. Another cause of cell loss is ischaemia, that is, as the tumour enlarges it tends to outgrow its blood supply. Ischaemia inhibits mitosis and causes the death of individual cells and groups of cells leading to the formation of a region of dead cells called the necrotic core. In some rapidly growing tumours, extensive necrosis is seen. We shall describe later the very important mechanisms that operate within a malignant tumour in order to enhance its blood supply and allow it to rapidly grow and invade surrounding tissue.

By the time a human cancer is discovered it has usually passed the initial phase of logarithmic growth and the rate has slowed down considerably due to cell loss. It follows therefore that growth rate alone cannot be used to estimate the age of a tumour.

3.3   The spread of tumours

Benign tumours proliferate locally and grow by expansion. They tend to compress the surrounding tissue, causing atrophy and the disappearance of its cells. The stroma of the surrounding tissue tends to be more resistant and becomes condensed to form a fibrous capsule around the tumour and this may increase in thickness as a result of desmoplastic reaction stimulated by the tumour. The edge of the tumour is consequently quite well defined. Clinical disorders produced by benign tumours arise mainly from mechanical effects such as obstruction of viscera or the pressure on nerves and organs. A benign tumour may remain in situ for years without causing ill effects and can be removed by surgery.

Malignant tumours grow by expansion and by infiltrating surrounding tissue. They are not encapsulated and their edges are often ill-defined. Projections of tumour cells extend from the central mass into surrounding tissues like the legs of a crab (the latin word for a crab is cancer). Their cells also invade the walls of the lymphatics and blood vessels in and around the tumour. The very aggressive behaviour of malignant tumours presents the major obstacle to their complete removal. Malignant tumours are usually poorly differentiated and grow rapidly, invade extensively and metastasise early. A key feature of the growth of malignant tumours is the phenomenon of angiogenesis which we consider in some detail later.

4 A Mathematical Model of Tumour Growth Based on the Theory of Nonlinear Elasticity

To begin our consideration of mathematical models of tumour development we concentrate on the early stages of growth and consider the tumour to be composed of a large, central necrotic core, surrounded by a layer of live proliferating cells on the tumour surface several cells thick. Initially the tumour is assumed to be spherical. All living tumour cells are assumed to be identical and each is considered to be an incompressible structure of fixed volume. The gross internal forces in the necrotic core are characterised by a pressure distribution P. Cell adhesion produces a surface tension force at the boundary. For a benign tumour where growth leads to a steady state there is a balance between nutrient supply and growth. Unstable development of the tumour arises when the internal pressure forces overcome the surface tension and inter-cellular adhesion. The instability is initially manifested as a pinching or a corrugation of the boundary surface at the equator of the tumour and more elaborate instability configurations may lead to further subdivision or disintegration with subsequent invasion of the surrounding host tissue. These ideas and assumptions form the basis of a model for tumour growth due to Greenspan [21].

The basic tool for our modelling of tumour growth is thin shell elasticity theory (see Landau & Lifschitz [34] and Fung [28]). We work in terms of averaged variables. That is we average over the reference thickness of the shell. Averaged variables will be denoted throughout by an over bar.

Let R be the initial radius of the tumour and r its current radius and denote by H and h the corresponding thickness of the layer of live proliferating cells. As a reference configuration we take $R = a$ at $t = 0$ and $h = H$ at $t = 0$.

With spherical coordinates θ and ϕ in the current configuration we take $\mathbf{a}_1$ and $\mathbf{a}_2$ to be the corresponding unit basis vectors and $\mathbf{a}_3$ to be the unit outward normal to the spherical tumour.

The average $\bar{\boldsymbol{\alpha}}$ of the deformation gradient $\boldsymbol{\alpha}$ (i.e. the measure of the deformation of the layer of variable cells) is expressed in terms of the above basis as

$$\bar{\boldsymbol{\alpha}} = \bar{\lambda}_1 \mathbf{a}_1 \otimes \mathbf{a}_1 + \bar{\lambda}_2 \mathbf{a}_2 \otimes \mathbf{a}_2 + \bar{\lambda}_3 \mathbf{a}_3 \otimes \mathbf{a}_3.$$

The averages $\bar{\lambda}_i$ of the principal stretches λ_i $(i = 1, 2, 3)$ are measures of how much the layer of live cells stretches or compresses along each of the (θ, ϕ, r)-directions corresponding to $i = 1$, 2 and 3 respectively. Under our incompressibility assumption we have

$$\bar{\lambda}_1 \bar{\lambda}_2 \bar{\lambda}_3 = 1,$$

and in view of the symmetry of the deformation we can write

$$\bar{\lambda}_1 = \bar{\lambda}_2 = \lambda \quad \text{and} \quad \bar{\lambda}_3 = \lambda^{-2}.$$

Consequently

$$r = \lambda R, \quad h = \lambda^{-2} H.$$

For a growing tumour $\lambda > 1$ and so as the tumour increases in size the cell layer decreases in thickness.

Let the average $\bar{\sigma}$ of the Cauchy stress tensor σ have principal components $\bar{\sigma}_{ii}$, $i = 1, 2, 3$. Then

$$\bar{\sigma}_{ii} = \bar{\sigma}_i - \bar{P}, \quad i = 1, 2, 3,$$

where $\bar{P}$ is the average of the hydrostatic pressure P and $\bar{\sigma}_i$ is the average of σ_i. For an incompressible, isotropic elastic material $\bar{\sigma}_i$ is given by (Haughton & Ogden [31])

$$\bar{\sigma}_i = \bar{\lambda}_i \frac{\partial \bar{W}}{\partial \bar{\lambda}_i},$$

where the "strain energy" function $\bar{W} = \bar{W}(\bar{\lambda}_1, \bar{\lambda}_2, \bar{\lambda}_3)$ is symmetric in its arguments. $\bar{W}$ is a measure of the elastic potential energy stored in the layer of live proliferating cells at the tumour surface.

In the present situation we apply the thin membrane approximation

$$\bar{\sigma}_{33} = \bar{\sigma}_3 - \bar{P} = 0.$$

In other words we think of the stress present in the tumour membrane as a generalisation of surface tension. It then follows that

$$\bar{\sigma}_{11} = \bar{\sigma}_{22} = \bar{\lambda}_1 \frac{\partial \bar{W}}{\partial \bar{\lambda}_1} - \bar{\lambda}_3 \frac{\partial \bar{W}}{\partial \bar{\lambda}_3} \equiv \frac{1}{2} \lambda \hat{W}_\lambda,$$

where

$$\hat{W}(\lambda) = \bar{W}(\lambda, \lambda, \lambda^{-2}) \quad \text{and} \quad \hat{W}_\lambda = \frac{d\hat{W}}{d\lambda}.$$

Next, from equilibrium considerations we have

$$P = \frac{h}{r} (\bar{\sigma}_{11} + \bar{\sigma}_{22}) = \frac{2h\bar{\sigma}_{11}}{r},$$

where P is the internal expansive pressure producing the inflation. So we see that in the spherically symmetric equilibrium configuration, the internal pressure is balanced by the product of the surface tension force (proportional to $\bar{\sigma}_{11}$) and the membrane curvature $1/r$.

Using the above equation we now have

$$P = \frac{H\hat{W}_\lambda}{R\lambda^2}.$$

The essential features of the initial phase of solid tumour growth are thus captured by the model, i.e. the gross internal forces are characterised by an internal expansive pressure distribution which is counter-balanced by a surface tension at the outer boundary of the tumour. Of course the cause of growth is the uptake of nutrients by the cells, while cells in the interior die and necrosis takes place here due to the lack of vital nutrients. We should note that the role of surface tension and surface curvature is not fully understood.

Suppose we now examine the effect of small perturbations on the stability of the tumour. We assume that a point $\mathbf{a}$ in the finitely-deformed, radially symmetric configuration is now given an incremental displacement $\dot{\mathbf{a}}$. Thus we have

$$\mathbf{a} \to \mathbf{a} + \dot{\mathbf{a}}$$

with unit base vectors $\mathbf{a}_1, \mathbf{a}_2, \mathbf{a}_3$ corresponding to the (θ, ϕ, r)-directions respectively. The incremental displacement is written as

$$\dot{\mathbf{a}} = v\mathbf{a}_1 + w\mathbf{a}_2 + u\mathbf{a}_3.$$

Assuming that the deformation of the membrane is axisymmetric (i.e. independent of ϕ) then the equilibrium equation corresponding to $i = 2$ leads to the solution

$$w = c\sin\theta.$$

Since this deformation does not alter the spherical shape of the membrane and hence the tumour we do not consider it further.

The equilibrium equations corresponding to $i = 1, 3$ then reduce to

$$(2\Sigma_2 - \Sigma_3)\, u_\theta + \Sigma_2 \left(v_{\theta\theta} + v_\theta \cot\theta\right) - \left(\Sigma_2 - \Sigma_3 + \Sigma_2 \cot^2\theta\right) v \;\; = \;\; 0,$$
$$(2\Sigma_2 - \Sigma_3)\left(v_\theta + v \cot\theta + 2u\right) - \Sigma_1 \left(u_{\theta\theta} + u_\theta \cot\theta + 2u\right) - r^2 h^{-1}\dot{P} \;\; = \;\; 0.$$

The Σ_i, $i = 1, 2, 3$ are the three independent elastic moduli of the tumour which may be calculated in terms of the components of the tumours instantaneous elastic moduli tensor (see Chaplain & Sleeman [17] for details). They can also be partially related to the strain energy function $\hat{W}$ as follows

$$\Sigma_1 = \frac{1}{2}\lambda\hat{W}_\lambda, \qquad (2\Sigma_2 + \Sigma_1 - \Sigma_3) = \frac{1}{2}\lambda^2 \hat{W}_{\theta\theta}.$$

The solution to the above incremental boundary value problem is unique at any stage along the considered stable path up to a critical configuration at which uniqueness fails and the solution path bifurcates producing abrupt changes in the growth of the tumour suggesting the onset of metastasis and invasion of the surrounding host tissue.

To solve the above equations we seek solutions of the form

$$u = \sum_{n=0}^{\infty} A_n P_n(\cos\theta),$$

$$v = \sum_{n=0}^{\infty} B_n \frac{d}{d\theta} P_n(\cos\theta),$$

and examine the bifurcating modes (i.e. non-trivial solutions) for different values of n. This leads to the following conclusions:

$n = 0$ corresponds to purely radial growth;
$n = 1$ gives rise to three cases:
    case 1 - translation with no deformation;
    case 2 - thickening or thining at the poles of the spherical tumour;
    case 3 - thickening or thinning of the upper and lower hemispheres;
$n = 2$ corresponds to a deformation of the surface to a peanut shape;
$n > 2$ leads to more complicated aspherical shapes suggestive of the
    onset of metastasis and invasion of host tissue.

Let us now consider the above discussion in relation to the strain-energy function W. We choose

$$W = \mu_r \phi(\alpha_r),$$

where

$$\phi(\alpha) = (\lambda_1^\alpha + \lambda_2^\alpha + \lambda_3^\alpha - 3)/\alpha$$

and μ_r, α_r are real positive constants and summation over a finite number of terms applied by the repetition of the index r. This form of W was introduced by Haughton & Ogden [32] in their study of rubber like materials. In the present context we have

$$\hat{W}(\lambda) = \mu_r \left(2\lambda^{\alpha_r} + \lambda^{-2\alpha_r} - 3\right)/\alpha_r$$

and this leads to

$$\Sigma_1 = \mu_r \left(\lambda^{\alpha_r} - \lambda^{-2\alpha_r}\right),$$
$$\Sigma_2 = \mu_r \left\{(\alpha_r - 1)\lambda^{\alpha_r} + (\alpha_r + 1)\lambda^{-2\alpha_r}\right\},$$
$$\Sigma_3 = \mu_r \alpha_r \lambda^{\alpha_r}.$$

On the basis of these forms of Σ_i, $i = 1, 2, 3$ and our bifurcation analysis we suggest the following criteria for assessing the degree of malignancy of a tumour:

benign tumour: $\alpha_r \in \left(-\frac{3}{2}, -1\right] \cup [2, 3)$,
malignant tumour: $\alpha_r \in (-1, 1)$.

These parameter ranges are obtained on the basis that for a benign tumour only the $n = 0$ mode exists whereas for a malignant tumour all modes exist. For a full discussion of these ideas see Chaplain & Sleeman [17]. For the use of elasticity theory in modelling other biological tissues see Demirary [27], Gou [29] and Bogen [13].

5   Tumour Angiogenesis

Solid tumours progress through two distinct phases of growth, namely the avascular phase and the vascular phase (Folkman [2]).

The transition from the dormant avascular state to the vascular state, wherein the tumour possesses the ability to invade surrounding tissue and metastasise to distant parts of the body, depends on its ability to induce new blood vessels from the surrounding tissue to sprout towards and then gradually to penetrate the tumour, thus providing it with an adequate blood supply and microcirculation.

To accomplish this neovascularisation it is now well established that tumours secrete a diffusible chemical substance known as Tumour Angiogenesis Factor (TAF) into the surrounding tissue and extra cellular matrix. Much work has been carried out into the nature of TAF and its effect on endothelial cells since the pioneering work of Folkman in the 1970's culminating recently in the purification of several angiogenic factors, the determination of their amino acid sequences and the cloning of their genes (see Folkman & Klagsbrun [3] and Paweletz & Knierim [7]).

Angiogenesis proceeds through essentially three distinct events. In the first event there is a rearrangement and migration of endothelial cells situated in nearby vessels. The main function of endothelial cells is in the lining of the different types of vessels such as venules and veins, arterioles and arteries, small lymphatic vessels and the thoracic duct. They form a single layer of flattened and extended cells and intercellular contacts are very tight. These cells are the principal actors in the drama of angiogenesis.

In response to the angiogenic stimulus, endothelial cells in neighbouring normal capillaries destroy their own basal lamina and proceed to migrate into the extra cellular matrix. Small capillary sprouts are formed by accumulation of endothelial cells recruited from the parent vessel. The sprouts grow in length by migration of the endothelial cells. At some distance from the tip of the sprout the endothelial cells divide and proliferate to contribute to the number of migrating cells. We can summarise these events as

(i) degradation of the basement membrane by enzymes secreted by the endothelial cells;

(ii) migration of endothelial cells;

(iii) proliferation of the endothelial cells.

For a detailed account of these events see the review of Paweletz & Knierim [7].

It is these three ingredients which are essential to the development of our mathematical model. Let $n(\mathbf{x}, t)$ denote the density of endothelial cells and $c(\mathbf{x}, t)$ denote the concentration of TAF in the host tissue at the point $\mathbf{x}$ and at time t. The model (see Chaplain & Stuart [18]) then takes the form:

$$
\begin{array}{ccccccc}
\text{rate of increase} & = & \text{diffusion} & - & \text{loss due to} & - & \text{decay of} \\
\text{of TAF} & & \text{of TAF} & & \text{cells} & & \text{chemical}
\end{array}
$$

$$
\begin{array}{ccccccc}
\text{rate of increase} & = & \text{cell} & + & \text{mitotic} & - & \text{cell} \\
\text{of cell density} & & \text{migration} & & \text{generation} & & \text{loss.}
\end{array}
$$

In mathematical terms we formalise the model for the TAF concentration as

$$
\frac{\partial c}{\partial t} = D_c \nabla^2 c - \frac{Qcn}{(K_m + c)n_0} - dc
$$

where D_c, Q, K_m, n_0 and d are positive constants. Here we have assumed that the local uptake of TAF by endothelial cells is governed by Michaelis-Menten kinetics. We also assume that the uptake rate depends linearly on the cell density, and that the chemical decay is linear.

To complete the specification we impose the initial condition

$$
c(\mathbf{x}, 0) = c_0(\mathbf{x})
$$

where $c_0(\mathbf{x})$ is a prescribed function chosen to describe qualitatively the profile of TAF in the external tissue when it reaches the limbal vessels. For boundary conditions we take

$$
\begin{aligned}
c(\mathbf{x}, t) &= c_b \quad \text{at the boundary of the tumour,} \\
c(\mathbf{x}, t) &= 0 \quad \text{at the limbus.}
\end{aligned}
$$

To model the endothelial cell density we begin with a general conservation equation of the form

$$
\frac{\partial n}{\partial t} + \nabla . \mathbf{J} = F(n)G(c) - H(n).
$$

where $\mathbf{J}$ is the cell flux, $F(n)$ and $H(n)$ are functions representing cell growth and cell loss respectively. We assume that mitosis is governed by a logistic law (Stokes & Lauffenburger [23]) by specifying

$$
F(n) = rn \left(1 - \frac{n}{n_0}\right),
$$

where r is a positive constant related to the maximum mitotic rate and n_0 is related to the carrying capacity. Cell loss is assumed to be a first order (i.e. linear) process and we write

$$H(n) = -k_p n,$$

where k_p is the proliferation rate constant. The initial response of endothelial cells to the angiogenic stimulus is one of migration. This is followed by cell proliferation. In order to incorporate this through the function $G(c)$ we assume that there is a threshold concentration of TAF below which cell proliferation does not occur. We choose

$$G(c) = \left\{ \begin{array}{ll} 0 & c \leq c^*, \\ (c - c^*)/c_b & c > c^*, \end{array} \right.$$

where $c^* \leq c_b$.

There is a substantial body of evidence (Ausprunk & Folkman [1]) that the response of the endothelial cells to the presence of the TAF is a chemotactic one. As a consequence, we assume that the flux $\mathbf{J}$ of endothelial cells consists of two parts, one representing random motion and the other chemotactic motion of the cells. Thus we write

$$\mathbf{J} = -D_n \nabla n + n \chi_0 \nabla c,$$

where D_n is the diffusion coefficient of the endothelial cells and χ_0 (assumed constant) is the chemical chemotactic coefficient. With these assumptions the equation governing the evolution of endothelial cells is

$$\frac{\partial n}{\partial t} = D_n \nabla^2 n - \chi_0 \nabla.(n \nabla c) + rn \left(1 - \frac{n}{n_0} \right) G(c) - k_p n.$$

We assume that initially the endothelial cell density is a constant n_0 say at the limbus and zero elsewhere in the extra cellular matrix (ECM), i.e.

$$n(\mathbf{x}, 0) = \left\{ \begin{array}{ll} n_0 & \text{at the limbus,} \\ 0 & \text{in the ECM.} \end{array} \right.$$

If we further assume that throughout the subsequent motion, the cell density remains constant at the limbus then we have

$$n(\mathbf{x}, t) = n_0 \quad \text{at the limbus.}$$

The boundary condition for the cell density at the site of the tumour is taken to be

$$n.\nabla n = 0.$$

From this assumption and our expression for the flux $\mathbf{J}$ we see that while n remains small this boundary condition approximates a zero flux condition at the tumour site. Biologically this means that the endothelial cells remain within the capillary sprouts. Thus as long as n remains close to zero the phase of the sprout moving across the ECM is modelled. However, when the tumour is reached, n is non-zero and so the flux $\mathbf{J} = n\psi_0\nabla c$ points into the tumour region. During this phase, the model is a crude approximation to the interactions between the endothelial and tumour cells which is important during the vascular phase of growth.

Under appropriate non-dimensionalising we finally arrive at the model system

$$\frac{\partial c}{\partial t} = \nabla^2 c - \frac{\alpha n c}{\alpha + c} - \lambda c,$$

$$\frac{\partial n}{\partial t} = D\nabla^2 n - \chi\nabla.(n\nabla c) + \mu n(1-n)G(c) - \beta n,$$

where

$$G(c) = \begin{cases} 0 & c \le c_*, \\ c - c_* & c_* < c, \end{cases}$$

together with the initial and boundary conditions

$$c(\mathbf{x}, 0) = c_0(\mathbf{x}),$$

$$c(\mathbf{x}, t) = \begin{cases} 1 & \text{at the tumour site,} \\ 0 & \text{at the limbus,} \end{cases}$$

$$n(\mathbf{x}, 0) = \begin{cases} 1 & \text{at the limbus,} \\ 0 & \text{at the tumour site,} \end{cases}$$

$$n(\mathbf{x}, t) = 1 \quad \text{at the limbus,}$$

$$n.\nabla n = 0 \quad \text{at the tumour site.}$$

Having developed the mathematical model, the next step is to test it against available data in a variety of situations to see whether the model predicts observed experimental results and to what extent it should be modified in the light of new evidence.

In addition it is important to obtain an understanding of the underlying and qualitative properties of our model. Such an investigation has been carried out in Chaplain, Giles, Sleeman & Jarvis [25] in the idealised one dimensional case, that is where the tumour is considered as a point mass situated at $x = 0$ and the limbus is the point $x = 1$.

In higher dimensions the problem is both analytically and numerically much more demanding and leads to a number of open problems both from the point of view of modelling as well as in the mathematical analysis.

Consider a two-dimensional problem of a tumour in the form of a circular disc of radius a implanted concentrically in a surrounding medium of radius b with the annulus $a < \rho < b$ being the ECM. For the TAF we take

$$c(\mathbf{x}, 0) = \frac{b^2 - \rho^2}{b^2 - a^2}, \quad a < \rho < b,$$

$$c(a, \theta, t) = 1, \quad c(b, \theta, t) = 0.$$

Here we are taking the usual polar coordinates $(x, y) = (\rho \cos \theta, \rho \sin \theta)$. In the numerical experiments we describe, the following parameter values were chosen on the basis of experimental evidence as

$$\alpha = 10, \quad \gamma = 1, \quad \lambda = 1, \quad D = 0.001, \quad \chi = 0.75,$$

$$\mu = 100, \quad \beta = 4, \quad c^* = 0.2, \quad a = 0.2, \quad b = 1.$$

The initial concentration profile of TAF is shown in Fig. 1. For the endothelial cell concentration n we imagine the initial configuration where there are six blocks of endothelial cells which will respond to the TAF as in Fig. 2. Figs. 3, 5 and 7 show the evolution of TAF while Figs. 4, 6 and 8 show the evolution and proliferation of endothelial cells towards the tumour. From Figs. 3, 5 and 7 we see that the evolution of TAF quickly attains a form not too different from its steady state. This suggests a further simplification to the modelling and leads to more tractable analysis.

In the one-dimensional case we take $c(\mathbf{x}, t) = c_0(x) = 1 - x$ for all t and in the two-dimensional case we take $c(x, t) = (b - \rho)/(b - a)$. In both cases the governing equation for n becomes decoupled and numerical simulations give results of the same qualitative form as before.

6   Conclusions

We began this article by describing the essential characteristics of solid tumours and demonstrated how the theory of nonlinear elasticity can be used to characterise the structure of solid tumours leading to mechanisms for distinguishing between benign and malignant tumours. The idea of a strain-energy function plays a central role in the development of the theory. This is only the beginning of thinking about tumours in this way. Further development must be directed towards obtaining a greater insight into the modelling of tumour-strain energy functions. This is an important problem and can only be solved in collaboration with medical scientists and practitioners. Furthermore our elastic theory is "static" and needs to take into account the dynamics of tumour growth.

The second part of the development concerns the role of angiogenesis and we have shown how this can be modelled through a non-standard reaction diffusion mechanism. In the idealised one-dimensional case a considerable amount of qualitative information can be obtained through nonlinear analysis. For the more realistic two and three dimensional situations there is much to be done analytically.

In addition to these problems there are a number of important issues which need to be addressed including

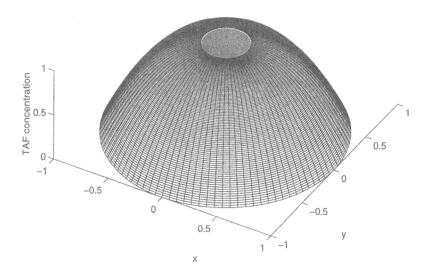

Figure 1: TAF concentration at $t = 0$.

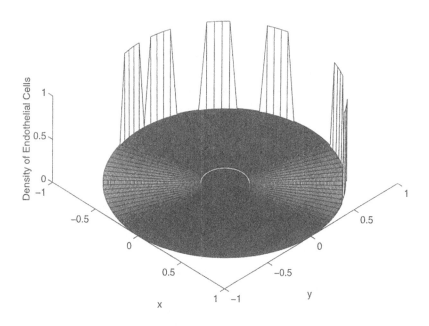

Figure 2: Endothelial cell distribution at $t = 0$.

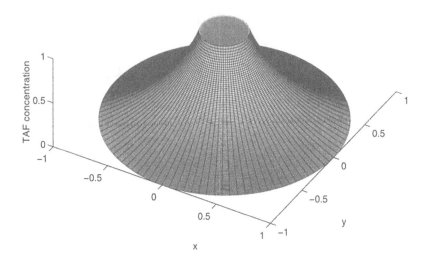

Figure 3: TAF concentration at $t = 0.25$.

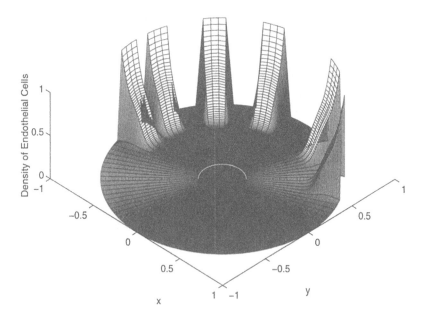

Figure 4: Endothelial cell distribution at $t = 0.25$.

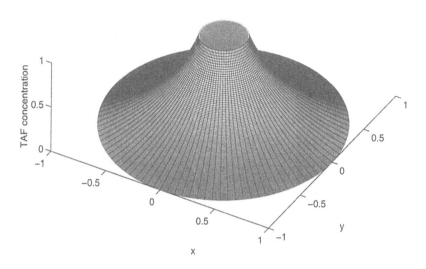

Figure 5: TAF concentration at $t = 0.6$.

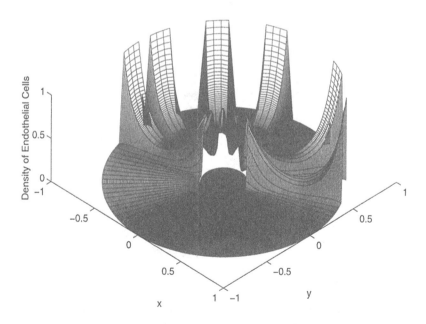

Figure 6: Endothelial cell distribution at $t = 0.6$.

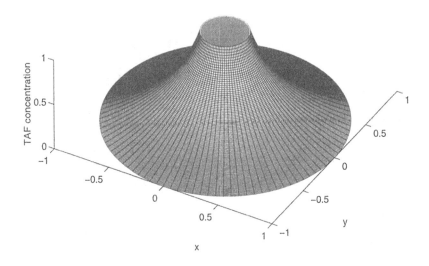

Figure 7: TAF concentration at $t = 1.0$.

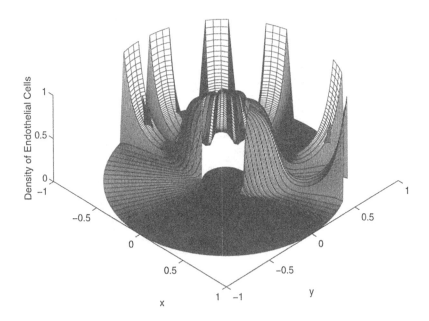

Figure 8: Endothelial cell distribution at $t = 1.0$.

- What happens when the angiogenic capillaries penetrate the tumour?

- How does the tumour respond under vascularisation?

- How is branching of capillaries achieved?

- Precisely how does the TAF interact with the endothelial cells of the limbus?

- How does the ECM affect diffusion?

These are all important questions and some of them are currently under investigation.

References

Tumour Properties

[1] Ausprunk, D.H. and Folkman, J., Migration and proliferation of endothelial cells in preformed and newly formed blood vessels during tumour angiogenesis. *Microvasc. Res.* **14**, 53-65, 1977.

[2] Folkman, J., The vascularisation of tumours. *Sci. Am.* **234**, 58-73, 1976.

[3] Folkman, J. and Klagsbrun, M., Angiogenic factors. *Science* **235**, 442-447, 1987.

[4] Folkman, J. and Moscona, A., The role of cell shape in growth control. *Nature* **273**, 345-349, 1978.

[5] Muir, R., In *Muir's Text Book of Pathology*, ed. J.R. Anderson, 12th Edition, E. Arnold, London, 1985.

[6] Perham, P.M., Robertson, A.J. and Brown, R.A., Morphometric analysis of breast carcinoma: association with survival. *J. Clin. Pathol.* **41**, 173-177, 1988.

[7] Paweletz, N. and Knierim, M., Tumour-related angiogenesis. *Crit. Rev. Oncal. Hematol.* **9**, 197-242, 1989.

[8] Sutherland, R.M., McCredie, J.A. and Inch, W.R., Growth of multicell spheroids in tissue culture as a model of nodular carcinomas. *J. Nat. Cancer. Inst.* **46**, 113-120, 1971.

[9] Thomlinson, R.H. and Grey, L.H., The histological structure of some human lung cancers and the possible implications for radiotherapy. *Br. J. Cancer* **9**, 539-549, 1955.

Modelling Tumour Growth

[10] Adam, J.A., A simplified mathematical model of tumour growth. *Math. Biosci.* **81**, 224-229, 1986.

[11] Aroesty, J., Lincoln, T., Shapio, N. and Boccia, G., Tumour growth and chemotherapy: mathematical methods, computer simulations and experimental foundations. *Math Biosci.* **17**, 243-300, 1973.

[12] Balding, D. and McElwain, D.L.S., A mathematical model of tumour-induced capillary growth. *J. Theor. Biol.* **114**, 53-73, 1985.

[13] Bogen, D.K., Strain energy description of biological swelling. I Single fluid compartment models. *ASME J. Biomech. Eng.* **109**, 252-256, 1987.

[14] Brzakovic, D., Luo, X.M. and Brzakovic, P., An approach to automated detection of tumours in mammograms. *IEEE Trans. Med. Imag.* **9**, 233-241, 1990.

[15] Burton, A.C., Rate of growth of solid tumours as a problem of diffusion. *Growth* **30**, 157-176, 1966.

[16] Chaplain, M.A.J. and Sleeman, B.D., A mathematical model for the production and secretion of tumour angiogenesis factor in tumours. *IMA J. Math. Appl. Med. Biol.* **7**, 93-108, 1990.

[17] Chaplain, M.A.J. and Sleeman, B.D., Modelling the growth of solid tumours and incorporating a method for their classification using nonlinear elasticity theory. *J. Math. Biol.* **31**, 431-473, 1993.

[18] Chaplain, M.A.J. and Stuart, A.M., A model mechanism for the chemotactic response of endothelial cells to tumour angiogenesis factor. *IMA J. Math. Appl. Med. Biol.* **10**, 149-168, 1993.

[19] Greenspan, H.P., Models for the growth of a solid tumour by diffusion. *Stud. Appl. Math.* **51**, 317-340, 1972.

[20] Greenspan, H.P., On the self-inhibited growth of cell cultures. *Growth* **38**, 81-97, 1974.

[21] Greenspan, H.P., On the growth and stability of cell cultures and solid tumours. *J. Theor. Biol.* **56**, 229-242, 1976.

[22] Gyllenberg, M. and Webb, G.F., Quiescence as an explanation of Gompertzian tumour growth. *Growth Dev. Aging* **53**, 25-33, 1989.

[23] Stokes, C.L. and Lauffenburger, D.A., Analysis of the roles of microvessel endothelial cell random motility and chemotaxis in angiogenesis. *J. Theor. Biol.* **152**, 377-403, 1991.

[24] Wright, N.A., Cell proliferation kinetics of the epidermis. In *Biochemistry and Physiology of the Skin*, ed. L.A. Goldsmith, 203-229, Oxford University Press, 1983.

Mathematical Analysis

[25] Chaplain, M.A.J., Giles, S.M., Sleeman, B.D. and Jarvis, R.J., A mathematical analysis of a model for tumour angiogenesis. *J. Math Biol.* **33**, 744-770, 1995.

[26] Cummings, F.W., On surface geometry coupled to morphogen. *J. Theor. Biol.* **137**, 215-219, 1989.

[27] Demirary, H., Large deformation analysis of some soft biological tissues. *ASME J. Biomech. Eng.* **103**, 73-78, 1981.

[28] Fung, Y.C., *Biomechanics*. Springer, New York, 1981.

[29] Gou, P.F., Strain energy functions for biological tissues. *J. Biomech.* **3**, 547-550, 1970.

[30] Hart, T.N. and Trainor, L.E.H., Geometrical aspects of surface morphogenesis. *J. Theor. Biol.* **138**, 271-296, 1989.

[31] Haughton, D.M. and Ogden, R.W., On the incremental equations in nonlinear elasticity. I Membrane theory. *J. Mech. Phys. Solids* **26**, 93-110, 1978.

[32] Haughton, D.M. and Ogden, R.W., On the incremental equations in nonlinear elasticity. II Bifurcation of pressurised shells. *J. Mech. Phys. Solids* **26**, 111-138, 1978.

[33] Jones, D.S. and Sleeman, B.D., *Differential Equations and Mathematical Biology*. George Allen & Unwin, London, 1983.

[34] Landau, L.D. and Lifschitz, E.M., *Theory of Elasticity*. Pergamon, London, 1959.

[35] Lu, G. and Sleeman, B.D., Maximum principles and comparison theorems for semilinear parabolic systems and their applications. *Proc. Roy. Soc. Ed. A* **123**, 857-885, 1993.

[36] Murray, J.D., *Mathematical Biology*. Springer-Verlag, New York, 1989.

[37] Protter, M.H. and Weinberger, H.F., *Maximum Principles in Differential Equations*. Springer-Verlag, New York, 1984.

[38] Todd, P.H., Gaussian curvature as a parameter of biological surface growth. *J. Theor. Biol.* **113**, 63-68, 1985.

Printed in the United States
By Bookmasters